WHAT READERS HAVE SAID ...

Reckoning
Discoveries after a Traumatic Near-Death Experience

"...an **insightful and uplifting** book ... destined to become an **authoritative classic** in the realm of near-death experience literature." **Mark Anthony** the Psychic Lawyer/Psychic Explorer® *(Evidence of Eternity and Never Letting Go)*

Dancing Past the Dark: Distressing Near-Death Experiences

"...**this brilliant book!** Put aside whatever you think you know about NDEs and read this. It is about the entire NDE experience as well as how it has been perceived, studied, contested, and reported, and so much more. I cannot recommend it enough." **Anne Rice** *(Interview with a Vampire)*

"**Absolutely enthralling–literary, adventurous, incisive, informative and smart.** One of the strongest, most thought-provoking books on the paranormal I've ever seen." **Steve Volk**, *(Fringe-ology)*

"*Dancing Past the Dark* is **the bible of its subject.**" **Bruce Greyson, MD**, a founder of IANDS, lead NDE researcher

The Buddha in Hell and Other Alarms

"Extremely well written...She really **put my mind at ease** in her explanation of hellish NDEs and it made total sense. I highly recommend this book." An Amazon **reviewer**

RECKONING

RECKONING

DISCOVERIES AFTER A TRAUMATIC NEAR-DEATH EXPERIENCE

NANCY EVANS BUSH, MA

Reckoning: Discoveries after a Traumatic Near-Death Experience
Copyright 2021 by Nancy Evans Bush.

All rights reserved. No part of this book may be used or reproduced in any manner whatsoever without written permission from the author, except for brief quotations in critical articles or reviews, and fair use in academic papers and educational materials.

9780985191733

Cover design by WebsiteGeographer, http://www.webgeographer.com.
Cover photograph of the grand design spiral galaxy Messier 100, credit NASA / ESA / Hubble.
Book design by Pressbooks.
The website http://dancingpastthedark.com gives access to blog posts, articles, and contact information for the author.

DEDICATION

This book is dedicated to Kenneth Ring, PhD, one of the "fathers of the field": psychologist, pioneer researcher and author, a founder and first president of the International Association for Near-Death Studies. Ken brought me in as IANDS' first hire (and did not boot me out when he discovered what came with me). He was the first to ask, "How have you made sense of your experience?" At the time, I had no sense to make and gave him a flippant reply. Thirty-some years on, Ken, here at last is my answer to the question we were both asking, now with much gratitude and a toast to unconditional love through tough times.

The undeniable thing about the Void is that it just won't go away. Nothing, no amount of faith or reason or denial, can budge it. Once it has 'known you', you are its for good. Thus, reconciliation seems the only viable way.
Steven Weber

But [after wrestling all night with the stranger], Jacob said, "I will not let you go unless you bless me."
Genesis 32.26

Yet the terror of the abyss is also a sky—the luminous void—for the integral human being.
Jean Gebser

Heaven and hell are within us, and all the gods are within us… All the gods, all the heavens, all the world, are within us.
Joseph Campbell

CONTENTS

	Preface	xiii
	PART I. WHAT WE BRING WITH US	
1.	Beginnings	3
2.	A Flipped Cosmology	25
3.	Rites of Passage	51
	PART II. WHAT WE DISCOVER	
4.	The 'Para' Realms	65
5.	Stages: the Unfolding Context	81
6.	Archetypes: Images and Forms	107
7.	A Renovated Structure	131
8.	Chaos and Fractal Being	145
	PART III. WHAT WE BRAID TOGETHER	
9.	Integration: Marinate in NDEs	171
10.	Temptation, Its Own Endless Dream	197

11.	Is This Real	199
12.	The Void: El Collie and strange places without answers	207
13.	The Map Is Not the Territory	223
	Acknowledgments	231
	Appendix: Recommended Reading	235
	Bibliography	237

PREFACE

Almost forty years ago, in that first decade of the field of near-death studies, I became the first paid staff person in the world's first nonprofit organization on the subject, the International Association for Near-Death Studies (IANDS). That was a remarkable spot from which to observe and participate in those early years when public excitement amounted almost to a pandemonium. None of us knew, at the interview, that I came as a near-death experiencer; indeed, I had never heard of NDEs before. But there was a horrifying experience which had been disturbing me for years, and now I would know its name.

At IANDS, because my experience was frightening rather than beautiful, it was an outlier, and I was not the only one frightened; many, it seemed, did not consider it a true NDE, and some would consider my presence a negative influence. Obviously, someone would have to look into the matter. And there I was, with the incoming mail and the ringing phones, able to build a small, slow collection of difficult experience accounts. I was driven to find information for my own questions but always aware of the others behind the letters and phone calls. Another inconvenience was that in an environment rich with paranormality, I had no easily identifiable psychic gifts, although a visitor who was introduced as a Hawaiian Kahuna later told me, "You do not have a specific gift because you are to be open to them all." That was kind but did not mean I understood any of it, including my own experience. It would take time to see those 'failings' as gifts.

The NDE had happened when I was 28. It brought no rapture, no peace; but neither was it traditionally hellish, though it ripped up my notions of reality. It was unusual and shattering, and could not be ignored. Like the beautiful NDEs, my experience lasted only moments, but processing it has taken the rest of my adult lifetime. For years people have been asking how I explained it. Decades after the NDE, my response has been universally evasive because although I had become a so-called "expert" about distressing NDEs in general, I still had no real grip on my own.

And then–here's a big time jump–the Covid 19 pandemic shut down life as we knew it, opening an immense sweep of time with no engagements, no meetings, no activities of any kind. Such a vast, unburdened time and silence allowed me to take on that long-ago experience which has so occupied my life and to wrestle with its questions and demand answers…and to write this book.

Given that background, the questions this book explores are somewhat different than those of the many autobiographical books about individual NDEs and the thoughtful, usually scientific, commentaries. These questions go beyond description of my NDE event to ask about the nature of the reality shaping it and all NDEs, and to examine how one makes sense of it all over time. Despite its personal anecdotes, the book is less about me than about the ideas which powered my near-death experience, reshaped my cosmology, and may have relevance for others. The circumstances have been such that the *process* seems to me worth sharing, and some aspects of the reconciliation will be interesting to those who have watched the field of near-death studies grow. Now, here is what may be my apologia for the approach of this book. The integrating process has swooped in repeatedly, not in rapturous feelings, except in flashes, but in Big Ideas of one sort or another, and with deep understandings. There is, to be honest, a lot of nerdiness in my integration of the NDE.

This is not the way I hear most NDEs being discussed. My feelings are deep but not effusive; the spirituality is deeper, often soundless. Once more, I am trekking off on a road so little traveled I know almost nothing in any detail about who else is on it. I know there are others, for I have taken phone calls and answered letters, and talked with people; the *dancingpastthedark* blog has readers who write to me; yet with few exceptions, we barely know each other as individuals.

So, the NDE itself was stylistccally something of a minimalist affair, and the integration has been likewise. Similarly, despite tumultuous emotional underpinnings, the visible tracks of integration are primarily intellectual rather than the prettiness of Hallmark. Universe said, go figure this out, and the answers have come in the form of Big Ideas. At least, Big for one who is not a mathematician. To understand what was going on, you will need enough explanation to know what I was seeing. I will do my best to keep the nerdy stuff as alive as I know it to be. What I can do is let you know what I was *feeling* as well as *learning*.

This is an account of following breadcrumbs and threads in a hunt for understanding. The process has taken decades. The hunt which set me on a lonesome path to find answers has led here, where my reckoning has brought us, and I am glad to know them at last, and to share them

PART I.

WHAT WE BRING WITH US

As presenter or president, every time I have stood at a microphone for a conference or radio or television interview, thinking of the audience settling into their seats, I have been aware of the tantalizing presence of deep stories. Today's enlarged and invisible audience at the far end of the lights and cameras now live-streaming an event worldwide brings even more stories. Although no two of our near-death or spiritually transformative experiences are identical, we share common patterns and similar questions, and we all filter answers through our own personal sagas. All those backgrounds, all those histories, those hopes and fears and expectations! As I scan the faces, I always wonder, "How did we all come to be here? What stories are we telling?" Here is a part of my story.

CHAPTER 1.

BEGINNINGS

MAPS

When my sister Barbara and I were growing up, the center of our bedroom floor was occupied by a tan linoleum rug, probably 6×8 feet, printed with a map of the United States. Each state capitol was marked by a star and a sketch of the state's principal product. I cannot remember a time before the map. That map taught us our first lessons in geography and commerce, and something of the romance of place. It established territory: "That river is on my side. *Don't you dare cross it!*" The rug was a treasure of our childhood. Kansas, I remember, was a muddy yellow, marked with a sheaf of wheat. Our mother was raised in a small town well to the left of the sheaf.

In the summer before I started kindergarten, when Barbara was three, our mother was determined to take us to visit her folks. She was an experienced driver because for ten long summers after her high school graduation she had been a pianist and dramatic reader with the Chautauqua educational tent show circuit, crisscrossing town to town across North America. She was not put off by the thought of driving on her own with two small girls from upstate New York to southwestern Kansas, unfazed by a distance substantially longer than that map on the floor of our bedroom.

The time was pre-World War II. Eisenhower's highways were not even a dream. Mother packed us into the back seat of the '39 Chevy and took off for the Main Streets of countless small towns and the sedate open roads with 45-mile-an-hour speed limits tracing the mind-numbing distances of America's agricultural heartland. For days, Babs and I counted white horses and red barns, looked at books and clouds, squabbled, and ground little bits of Crayola into the scratchy upholstery. Lying on the floor or the stiff back seat, we watched phone wires swoop and swoop beyond the car windows, and we whined about the seeming endlessness of it all. America went on, tediously, *forever.*

We were experiencing the truth that a bedroom map is not the territory it represents.

It is an even longer way between a childhood rug and an adult near-death experience, but I often think of that map. A lifetime later, I have occasionally been asked, "What do you *really* think about near-death experiences—do you believe them? Are they real" After driving that interminable, visibly undivided distance, did I believe New York? Or Ohio or the everlastingness of Indiana corn? Or Kansas? I admire the response given by the great anthropologist Margaret Mead when asked long ago whether she believed in U.F.O.s. Writing in Redbook in 1974, she called it "a silly question":

> "Belief has to do with matters of faith; it has nothing to do with the kind of knowledge that is based on scientific inquiry. ... When we want to understand something strange, something previously unknown to anyone, we have to begin with an entirely different set of questions. What is it? How does it work?"[i]

Those have been my questions, and they shape this book. They mark the difference between belief, which is the map, and evidence, the territory. The questions help determine what is real. The dividing lines between those states are invisible, as are our experiences; yet we believe the states are real. Our experiences are invisible, hidden within the sensory and cognitive systems of individual, real experiences though not visible and not directly sharable. What are they? How do they work? How does the world work?

Whether a real experience is also factual is another question. The great majority of folk, hungry for some reassurance that life has meaning, long to believe *something* but aren't quite certain what is genuinely credible about NDEs and experiences like them.

Before attaching ourselves irretrievably to any explanation, we might pause to recall the old story of the blind men who blundered into an elephant, each encountering a different body part. When they tried to describe it, each according to his own experience of the animal, the argument went on and on, each of them partially right as much as each was wrong. And the elephant went its enormous way, a vast territory imperfectly described.

Like the quarreling blind men, none of us can see clearly enough to define all of reality, yet we attach ourselves to the part we grasp as if it were the whole. Like a little girl's first understanding of territory described by a linoleum rug, our individual interpretations are the maps of our own life experience, having only a sliver of correspondence to the enormity of the whole.

This book, then, is a collection of crumbs and threads and maps gathered as I worked to understand the single most influential experience of my life. The destination is an enormity.

WHO IS SPEAKING

I was born at home—well, at my maternal grandparents' temporary farmhouse in the minuscule town of Humansville, Missouri, in the Ozarks. It was the midst of the Great Depression, and we were there because the adults were in Depression joblessness. My grandfather had sold his Kansas newspaper for lack of business, and my parents had lost their jobs with touring companies on the Chautauqua circuit, because that cross between inspirational Vaudeville, circus tent, and educational television had been overtaken by radio. My father was serving as pastor of a small church in a neighboring town.

The Missouri stay would be short. Neither of my parents was cut out for rural life. My mother had been with Chautauqua for ten years, as a pianist who also did dramatic readings, and my father, twenty years her senior, had been a popular humorist and inspirational speaker. Shortly after my first birthday we were in Rochester, New York, with a larger church and city lights. And a baby sister.

Church had been my father's family business for generations, dating back to an ancestor who translated the Bible into Welsh; it is still considered to be for Wales what the King James is for English. Our grandfather's reputation as a championship preacher in North Wales brought him to this country to serve as an editor of the new *International Standard Bible Encyclopedia*, which, in digital format today is still in active use. Our father, tall and good looking, both funny and scholarly, had Welsh as his first language, with English his second. grown up in Wales, As an adult in the United States, he had been well paid as a lecturer with Chautauqua. Until the end of his life, he was periodically booked for speeches around North America. Even though the Depression had made our family, quite literally, poor as church mice, we knew we could hold our heads up.

Now all that family history had its chief office in our front living room, which served as the main conference room for the church, seconded by the guest room, where missionaries of many skin tones and dispositions came and went during those years around World War II. When those guests were not present, it was God's room, and we were not to mess with it.

During the week, our father's suits were on the thin and shiny side; but every Sunday morning he donned his formal morning dress—grey trousers with a thin black stripe, a single-breasted grey waistcoat with his father's gold watch and chain, and his formal morning coat (black, with swallowtails)—and walked, with his pretty wife and starched little girls, to church to praise God and preach to the working-class congregation who knew, if only by his dress and accent, that they were led by a man of God.

Theologically, we were in a Congregational church—liberal, which meant it held loose ties to the rigidity of creeds and bishops. We lived in service of an authoritative yet loving God the Father (whom Jesus called 'Abba' or papa), with a deep love of Jesus and his teachings in the Gospels, and an all-pervasive Holy Spirit. Our task was to love God and others and be of service to them, as Jesus taught. The wrathful God and bleak, guilt-ridden theology of Reformation thinkers was a part of Protestant history, so we knew about it vaguely, but anything my sister and I learned about hell came from the Christian Missionary Alliance Tabernacle across the street, whose enthusiastically fundamentalist Daily Vacation Bible School we were permitted to attend because our church didn't have such a program. (It taught us colorful, even gruesome, hymns unlike the sweetly pastel songs of our more free-thinking Sunday School, and we learned about religious terror.) So my sister and I were marinated in God and service, and also shaped by World War II.

Rochester, with Eastman Kodak as its primary industry, was a war center. Every child in our neighborhood—and there were a lot of them—had at least an uncle or a cousin, sometimes even a father, serving in the military, and my friend Edith, at the end of our block, had escaped with her parents from somewhere in Germany. She had a number tattooed on her arm. To this day, I have a visceral reaction to swastikas or anything else reminiscent of the Gestapo, a response as indelible as Edith's tattoo.

Primarily we were encouraged to lives of service, for which we were not too young. By the time I was six, a neighbor lady, head of the local Women's Christian Temperance Union, had organized our neighborhood gang into a branch of the Loyal Temperance Legion. (The first slogan of the national LTL was, 'Tremble, King Alcohol. We Shall Grow Up!') We had actual meetings at which she taught us about the Evils of Drink and the goals of Clean Living and Saving Souls. We scoured the nearby sidewalks and curbs to clear them of cigarette butts, cleaned our part of the street after the ice man's horse, and Babs and I got into trouble when we found a bottle of beer in the family refrigerator and poured it down the sink to save our father's soul.

At eight I organized a march of the LTL—the bravest thing I have ever done, for it involved escaping maternal supervision and venturing four city blocks away, into a commercial area. We marched, about six of us Legioners, armed only with shiny brown chestnuts (in case of need), past the witch's house, across the double-wide boulevard, over the bridge across the forbidden trolley tracks, and along the solid brick side wall of the saloon which was our destination. We had been told about the Devil's work there, how he tempted the unwary and taking their souls. Some adults might have said the unwary were Kodak workers stopping for a beer after their shift, but we knew the truth of the Devil's depravity and their need for rescue. We were so scared! But we strengthened our resolve by thinking of their sad wives and children, their pitiful lives being destroyed by the drunkenness of their men inside this building, who needed to be freed from Satan himself. We marched all that way, then summoned our courage, and a little Legioners posse burst through the swinging doors, screamed "Repent and be saved!" and we ran for our lives before the Devil could catch us.

As I said, it was the bravest thing any of us has probably ever done. The Congregational part of our lives would have cheered the caring and activism while thinking our cause a bit over-stated. As it was, we kept our mothers in the dark about it.

The cruelest effect of the war on me was that my best friend, Carol Grace, died when we were seven. She had been hospitalized for a kidney infection, but treatment was limited as the new miracle drug, penicillin, was reserved for the troops.

We were a family that loved dolls, but Carol had never had a nice one. Knowing this, and knowing, as I did not, that she was dying, my father arrived at one of his pastoral visits to her in the hospital, bearing every little girl's dream: a bride doll lavish in white tulle and rustly skirts. I had a single glimpse of the doll and her wedding finery and remember that stab of envy, but I knew how much Carol Grace would love her. The doll was so much loved over the next, last, week, they were buried together. It was a comfort to know that Carol Grace was not lonely, and when I got to heaven we would play with the bride doll together.

Along with that first experience of grief, I was marked by the kindness of my father's gesture, in a time of general desperation, with his own family always close to poverty, translating so much into an extravagant and perfect gift. He and I would go on to have a rocky history, but this story has always offered a measure of balance about his truer self.

By my teens we were in White Plains, thirty-some miles north of New York City. My father and I were in serious difficulty. He was twenty-three years older than our mother, and the demands of family life were wearing on him. Further, that bottle of beer in Rochester had been a clue, and he was drinking regularly and heavily, cheap jug wine stashed in the bottom file drawer of his study. He could be up until all hours night after night, most of the night, practicing his sermons but building a rage he could never shake: his failure to achieve the kind of church success he had been expected to attain; his loss of financial solidity after Chautauqua; the burden of by now four daughters, an endless drain on his energy and funds, not to mention his patience. None of that was ever said, of course, but he would lumber through the house night after night for days at a time, mumbling sermons, shouting drunkenly, banging doors and growling and name-calling. It kept us emotionally battered and sleep-deprived but the next day protecting his public image with our silence.

And of course there was still the parish with its complaints and demands. That involved all of us. There was not only Sunday worship but the meetings coming through the living room, more meetings and choir practice at church, the weekly folding of Sunday bulletins, helping with Sunday School, youth group, summer conferences, Women's Group, Men's Group, church potluck suppers.

I walked into town and joined the NAACP because I felt they should know someone cared, and besides, it was not church. They were very kind and let me volunteer occasionally by folding brochures. Did I believe all the church teachings at home? Certainly yes, most of it, and surely believed it was wrong of me not to have more faith in the rest; but between the theology and my father's behavior when drinking ranged a deep confusion, and at least I knew for a certainty that I believed in the NAACP.

The theological issues widened, of course, in college. First it was the shock of finding Christianity referred to in the history text as "a small sect." That really shook me. One thing led to another, and unanswerable questions piled high. I braved Mass with friends and fell in love with the candles and statuary but not the dogma; and observed Shabbos with other friends and fell in love with the sense of community and going to temple like Jesus but couldn't let go of Christmas; and felt generally adrift for a long time. The doubts did not keep me from excitement about the Billy Graham Crusade coming for the first time to New York City, and Barbara and I went to both Madison Square Garden and Yankee Stadium, and, yes, were 'saved' both times.

So that is the abbreviated background. I was a serious young person feeling over-responsible and under-performing; obviously bright, I had outstanding grades only in English and was routinely in summer school making up failed classes. I was emotionally fragile and probably depressed, but active in extra-curriculars. Behaviorally, my sins were many but neither lethal nor, I think, unforgivable: procrastination, meanness to little sisters and the girl in seventh grade gym I disliked intensely enough to put a fistful of dead flies down her back; a stolen blouse which I adored but could never wear because of guilt. Surely not enough to warrant God's eternal disdain! I have always believed in God as I understood the concept, I deeply wished Mary were Protestant, and I have loved the notion of Jesus but could not credit everything I was supposed to believe about him. Was that enough?

AN AUSTRALOPITHECINE SYNCHRONICITY

A foreshadowing synchronicity in my NDE story came early on, with no reason to suspect any prescience; it's more a pre-synchronicity. In the late 1950s, in New York City fresh from dropping out of my first graduate study, I was inexplicably hired as librarian for a prestigious foundation for anthropological research in a glorious townhouse on the Upper East Side of Manhattan. The hiring was inexplicable because, first, I was in no way a librarian, and second, I was equally free of any background in anthropology; however, I was a voracious reader and, at the time, fluent in French.

It was a boutique foundation with high standards in an elegant neighborhood two doors from Fifth Avenue. The Italianate building's street level held offices and kitchen with a dining area; second floor, the great library and conference rooms; third floor, several laboratories, including some mysterious and important *australopithecine* study (the foundation's specialty at that time was Early Man in Africa). The top floor was private, the director's residence.

A European aristocrat by birth, he became a prosperous gynecologist who moved to Hollywood and gained some renown as a director of award-winning films before persuading the originator of the foundation that its focus should be anthropology. It was largely his genius (and his assistant's) that shaped the small foundation into a force which in turn helped shape American anthropology for decades.

He was a force. He would sometimes, early in the morning, descend in his brocade dressing gown to the great library, throw open the French doors, and step out to the balcony for a breath of what so close to Fifth Avenue passed for fresh air. If we were fortunate, the Gabor sister whose equally splendid house was across the street would step out on her own balcony, and we would hear the two of them shouting and cursing at each other. They shouted in Hungarian, so it was unclear just what they were shouting about, but it was never dull.

Every day the staff gathered for a communal lunch around a long table next to the kitchen; there were perhaps a dozen of us, including the director at the head of the table, conducting conversation as if it were a performance. I regret not having taken notes. There was a session about pepper, the condiment. What did we know about pepper, and as our lack became obvious, he filled the gaps. Different kinds of pepper, how they were different, how they were harvested, what difference it made in their various communities. Who knew pepper could be so interesting? There was a session on furs, which one would each of us choose, and what difference would that make? His highest regard went to the staffer who said sable rather than mink.

And there was the session about a trip to Amazon, and his time with the chief of a cannibal tribe who asked what the visitor's people did with their dead. Told they were buried, the chief was horrified. "Buried? Buried?" And he became enraged, and spat. "Such waste!" The director thought it best to leave the area. Safe on 72nd Street, he was determined his staff should get it: Everything you can know about people is in their perspective.

My part of the library was, mercifully considering my ignorance, not the upstairs actual research area, which was overseen by the director's assistant. My part was in the basement, where some eight thousand uncatalogued books in thirteen languages remained mostly in storage boxes. Over the next two years, it would be my objective to unbox and shelve as many of them as possible, which meant I had to read enough to get an idea of where each belonged. And when the book was in a language I did not know, which was roughly half of them—well, go slowly, figure it out. Six hours a day of reading time, plus lunch with the director who thought it was his responsibility that his entire staff understand about differing perspectives.

That would be my unstructured but intensive introduction to anthropology, including elements of ethnography, archaeology, linguistics, folklorics, and shamanism, with various looks at the tools of exploring antiquity. I was expert at nothing, but stumbling my way through book after book provided a superb exposure to multiplicities of human beliefs and behaviors. What's more, we were surrounded by stars: not only the director and the Gabor sister across the street, but an Australopithecus was on the third floor, and at the quarterly cocktail parties for anthropological dignitaries (staff attendance required), I not only met Margaret Mead but learned that cultural anthropologists drank Manhattans and physical anthropologists and archaeologists preferred martinis; surely that was a clue to some valuable insight!

Off topic of NDEs, a favorite memory of that time involves a young woman who, with her husband, had fled Communist Hungary a year earlier, when the Soviets crushed the rebellion of 1956. She was now working in the upstairs library, where we were talking one day about the impending visit of Queen Elizabeth II to New York City.

"Oh, good," I said, "they'll be coming right up Fifth Avenue, and we can go see her."

She was incredulous. A real queen? And just anybody could go see her? *A real queen?* To the little girl who had grown up in a Communist country dreaming of princesses, this was too much to believe. But it was true. And on a Tuesday afternoon, Marjana and I joined a welcoming crowd lining upper Fifth Avenue, almost as many people as for the Macy's parade on Thanksgiving.

We ran down to about 68th Street for a breathless wait, and when President Eisenhower's bubble-top limousine began to pass us and we could see the small woman waving inside, we began to run north with her, Marjana keeping "her queen" in sight and waving until breath and legs would let us run no farther. I have often wondered if the queen was curious about the two young women who ran a half-mile and more alongside the limo with such determined and happy intensity. A wonderful moment, and a powerful lesson of what freedom means to those who have not had it.

MY NDE, 1962

[This account has a bit more detail about what I was experiencing of the Void than the version published earlier.]

It was a clear, hot night in late July. In an old Hudson River town of New York State, I was in labor with . *I have clarified a few de*my second child. Three weeks before the due date, premature labor had moved the baby into the birth canal, leading the obstetrician to order an emergency induction. Now, ready to deliver, I was anesthetized according to common practice.

What I knew next was that I found myself awake and somehow flying over a building. A quick glimpse backward—oddly, with no sense of turning around—and I could see unidentifiable box-like structures on the roof of what I thought must be the hospital, because there, up the hill, was the window of the classroom where I taught. There was the town, receding below me, and then the dark outline of hills along the Hudson, and the earth's curvature ("It really is round!"). Years later, I would describe it as hurtling into space "like an astronaut without a capsule."

The speed was puzzling. It felt like drifting but covering enormous distances at what seemed to be an angle, headed northeast. (Is there a northeast in space?) The nighttime darkness turned into immensity and a different sort of dark: it was "thinner" somehow, not opaque, shading inexplicably toward what might have been a paler horizon—except that there was no horizon, only a patch of brightness. My impression was that God was over there.

There was a sense of form to me, I recall, or at least of presence but no body—as if I were made of veiling—just insubstantial. But I was thinking. Did I *have* a mind, or was I *being* a mind? An unanswerable question.

The speed slowed or stopped, even. Perhaps a half-dozen circles appeared ahead and slightly to my left, half black and half white. There was no setting, no roadway, no landscape, and they simply appeared in the space as if on a minimalist set, clicking as they moved toward me, white-to-black, black-to-white. It seemed odd but not alarming until I realized they were sending a wordless but authoritative message. Their thought came all at once, all of it, without crowding:

"You are not real. You never were real. None of that [life] ever happened. This is all there is. This is all there ever was. This is It. Anything else you remember is a joke; it was all a joke. You are not real. Your life never existed. The world never existed. It was a game you were allowed to invent. There was never anything, or anyone.

They were simply reporting a business fact, not judgmental except a touch about knowing more than I did. I argued intensely, throwing out details of my mother's girlhood, stories of my husband's youth, facts from history—how would I know these things if someone had not told me? I couldn't have made up all those facts, all those languages and grammars—and books! No way could I have written all the books! And my first baby, the toddler waiting at home—she was real! And childbirth! What woman (even an imaginary woman) would invent childbirth?

"Whatever you remember, it's part of the joke. Your mother, your babies, the books—they were never real," they mocked ("We know, you don't.") "This is all there ever was. Just this."

But God? The thin darkness stretched off into nothingness, and the circles kept clicking.

And then whatever there was of me was entirely alone. But if I was unreal, what or who was reacting? I had no materiality, a wisp of consciousness. The circles had moved out of sight, and there was nothing left—the world unreal and gone, and with it everyone I knew and loved (but how had I known them, if they were never real?), and hills, and grass, and robins…it was the third act of "Our Town," but without even a self to go home to. I thought no one could bear so much grief, yet there seemed no end to it and no way out.

Everyone, everything, still gone, unreal, even the planet, even God, and I seemed to be dissolving into the twilight dark. Was this loneliness God? If I was unreal, how was so much pain real? Grief was there in my stead. I was not standing, for there was no one and nowhere to stand; I was merely being, for "there" did not exist, either. If the circles were speaking truth, and I was not real, who (?) was this?

There was no sense of hell or punishment, and though the circles might have had a touch of the smart-alec about them, their tone was primarily a clear-cut factuality. In style the whole first scene was minimalist, such vastness with Nothing.

Now, in the second scene, the light level was still like twilight, with darkness farther off, except for the one patch of brightness over where the horizon might have been, except there was no horizon. It was as if the air was kind of blue-grey, as if there were some light just behind it, which couldn't be seen but was there. There was sight, but there was nothing *to* see. No objects, only a vastness, but not a feeling of empty, either. I have no idea what that means; totally indescribable. Maybe like a faint engine, very far off, that you can't quite hear? The only feeling I had was the not-being loneliness, and the loss. Not only my loss, all loss. Rocks mourning, and moose in the woods, and whales, and seasons. *Everything. Every.Thing.* The Universe grieving for itself. Pure enormity.

~ ~ ~

And then I was groggily coming to in a hospital bed. *What was that?* My first waking thought: "Calvin was right. Predestination, and I must be one of the lost." And behind that, I knew a terrible secret presenting not in words but in knowing: *That is what is out there, what it will be like when I die, when we all die. This is simply a message. There is something so wrong with my very being, even God has willed me not to be.* (So then, who was that who is remembering being in the Void?)

I have a foggy recollection of being shaken or slapped awake and wanting to cry for help, but it did not last and I could not speak; I was immediately back in the wherever. My memory of the narrative contains no interruption, and I awoke directly from the Void.

A moment ago I was thinking, as there was no sense of personal judgment or punishment, why was my post-experience shock so great for so long afterwards? -–years and years! Also, why didn't I call God? Well, by then God was gone, too; but before, why didn't I? And the answer I get is, "It wasn't my place." Enough of answers with complications!

AFTER THE EXPERIENCE

The baby had been born cyanotic and was in intensive care; I was not allowed to see her. Was she real? I felt for her down the hall. Nothing was said about any medical event during delivery, and I did not ask. Nothing was real. No one asked if I had experienced anything unusual. My husband was there, and my mother, both worried about my distraught state, which I could not explain.

The hospital released me early because I was so distressed—the baby's uncertain condition, they thought. After some days the baby came home, so she and I met at last. I remember feeling glad to have her nearby, real or not. Now it seemed there were two little ones. At night, hearing a cry, I wondered, Should I get up? How can so much tiredness exist in a person who does not exist? If they are not real babies, do they need to be fed?

Psychic shock . Questions blanked, my only reaction to the circles became hyperarousal and instant shutdown. Only the shock was left, and the loss...of self, of everyone else, the loss of Creation, Earth, the holocaust, the enormity, the unspeakable aloneness. What did it all mean? As time passed, what might or might not be life moved on, seeming easier when I slammed shut any thinking about the message. Actual or not, little girls were fed and rocked and changed and tended to. Beneath the thinnest of emotional shells, despair ran roughshod. God had no place for me; the circles waited; nothing was real. Predestination, like a concrete block to the spirit: Betrayal of Abba and Jesus, replaced by an absent and ugly God. What chance do we have? But if we're not real, what if *that's* not real. If God isn't real...if the world isn't real...if I can't trust the Bible to be real (they have tricked my scholarly father and grandfather in their careful morning dress, and I am so angry)...what holds the world together—or is it together? So...death will be just that? How can an ordinary person know this? Too much.

It did not occur to me that I had been dead, though I thought of the scene as what I could expect when I *did* die; "dead" was a permanent condition, after all, and I was back and therefore not-dead. Neither did I consider it hell. Whatever hell might be, it was not that.

I tried once to tell my husband about the experience but stopped. Who can love a person and want to describe something so threatening, so full of grief? What would I say? I would not speak of it again for twenty years.

Six years went by, and one afternoon I went to have coffee with a friend. Heading into the kitchen, she gestured to a book on the table.

"Jung's *Man and His Symbols*. It just arrived. Take a look."

The book was large, profusely illustrated, something about images, and I leafed through it with interest. But then, from a left-hand page, one of the circles stared back. I froze. They were true! Someone else knew about the circles! In a storm of terror I hurled the book across the room and fled from the house, too frightened even to say goodbye, running hard the block and a half to my house, slamming through the heavy front door and locking it behind me. (Twenty-five years later, the friend would laugh and say, "Yes, I did think it odd that you simply disappeared.")

It would be several years more before I discovered that my "circle" was the Yin-Yang. I had been I troubled by the experience itself, but now—how does an unrecognized ancient Chinese symbol became a message-bearer in the experience of a mainstream Protestant in New England? (Critics say, 'But you could tell that you were real; why didn't you challenge them?' Because they were Absolute Knowers, Apparatchik; they belonged, I didn't.) Now, the Yin-Yang made the whole thing weird as well.

TWENTY-YEAR SHUTDOWN.

For twenty years I would tell no one about the experience. The early years were a theological tumult, as I raged at God for lying, and left church, only to go back for another try. I did not at any point try calmly to understand the experience, for which I simply had neither context nor vocabulary, though I flatly refused the idea of predestination. There were years of low-grade functioning and intervals of intense depression. Unfortunately, there was no budget for therapy; the approach had to be "do the best you can." Thank goodness for repression, which was generally my recourse.

Fourteen years after the NDE, my husband and I divorced, for reasons only marginally associated with my experience. The divorce was psychologically, financially, and socially devastating and personally shattering. With three children (two of them teens) and uncertain child support, I had to be able to support the four of us and within a short time was heading client services for an inner-city Federal employment and training program with 2,900 multi-cultural clients from a half-dozen sub-agencies. It was meaningful but insanely stressful work I was in poor condition to manage.

For the next four years, only a kind of manic will kept me operating at all, while the stresses kept accumulating—lingering issues of the NDE; the massively traumatic divorce and relational ill health; three kids in emotional free-fall and needing a stable mom who could not be there; my widowed mother retired to live nearby; having to sell my dream house in our "home" neighborhood; legal issues; financial issues; the intensity at work of client needs, relentless hardball office politics and misogyny, and my own failings at handling them all.

I wailed to a therapist, "I can't cope!"

He responded, "Look at it this way: You're at an intersection where four Mack trucks have collided, and you're lying under all of them. If you were coping well, I would have serious questions about your mental health."

After four years, I collapsed. All my old labels were gone and I didn't know who I was anymore: marriage, career, social standing, house, neighborhood, coping skills… all gone. My girls, equally overwhelmed, had escaped in their own ways — one to England, one to an early and traumatic marriage. I fled with my son, then age twelve, to the country, moving in with a supportive friend who saved us all and became my life partner.

PIVOT

Two years of stark depression and joblessness later, I had a teenage son at home and was running out of money. Any ad for high-level office work brought on shakes. What could I do? Then the synchronicities began: helping my now-married younger daughter hunt for a job, I came upon a small ad for a temporary position as office manager for a new nonprofit doing something I had never heard of a short drive from my home. My daughter said urgently,

"Mom, you should answer this!"

I did, and less than a week later, with considerable apprehension, I began work on the campus of the University of Connecticut as '"office manager" of the International Association for Near-Death Studies. IANDS.

The office was a 10' x 12' box containing a desk, a table, a couple of wooden chairs, and a single well-used file cabinet. My first morning there, two nervous students looked at the new hire, who was the age of their mothers, and asked awkwardly,

"Would you like to…uh…look at the files?"

Yes, I agreed, that would be a good place to start. Using two hands because it was heavy, one volunteer wrenched open a drawer so stuffed with correspondence that envelopes spilled over the sides. The second drawer was just like it, and the third—all of them jammed with letters which had been opened, read, sometimes answered, and always folded back into the envelope and crowded into a drawer, unalphabetized. All those people, breaking their silence, some of them after decades, hoping to hear something back. I looked at the file cabinet and the smiling students.

"Yes," I thought," I can do this job."

IANDS was housed in the Psychology Department, where the Association's president, Kenneth Ring, was a professor. He had a reputation in the department for odd ideas, as he was one of a tiny group of researchers in the U.S. inquiring about the academically suspect subject of near-death experiences. He had published the first scientifically-minded study of near-death experiencer demographics. Ken's instructions to me that first morning were simplicity itself. He looked at a shelf of books, gestured toward the file cabinet with its burden of first-person accounts, and—oh, genius!—said, "For the first three weeks, just read."

And so it began. Within those first weeks, my experience found its name. I still could not look the experience in the eye, and I had no idea there was such a thing as integrating an event like that—but those three weeks of introduction to NDEs were recognizably a pivot point. I had walked into a new life.

---

[1] Mead, Margaret. https://www.Nytimes.com/2020/07/28/Insider/ Ufo.

CHAPTER 2.

A FLIPPED COSMOLOGY

Sometimes a period of time seems stuffed with significance. In his prodigious book *Cosmos and Psyche,* cultural historian Richard Tarnas writes:

> Within the timespan of a single generation surrounding the year 1500, Leonardo, Michaelangelo, and Raphael created their many masterworks of the Late Renaissance…; Columbus sailed west and reached America, Vasco de Gama sailed east and reached India, and the Magellan expedition circumnavigated the globe, opening the world forever to itself; Luther posted his theses on the door of the Wittenberg castle church and began the enormous convulsion of Europe and the Western psyche called the Reformation; and Copernicus conceived the heliocentric theory and began the even more momentous Scientific Revolution. From this instant, the human self, the known world, the cosmos, heaven, and earth were all radically and irrevocably transformed. All this happened within a period of time briefer than that which has passed since Woodstock and the Moon landing.[i]

Cosmology's end

That is, to my mind, the Great Unhooking, a transition when Western civilization was forcibly detached from its ancient certainties. What must it have been like for ordinary people, for priests, for scholars, for sailors, to hear there was no end of the earth to sail off of, and the sun did not revolve around Earth, which was not the center of the entire universe?

"No! I don't believe it! Old fool, that one, messing with the stars."

Everyone knew humanity had been made a little below the angels; after all, wasn't Earth the center of all the stars? It's always been Earth at the heart of the whole heavenly cascade, so how could anyone believe it was merely a second-rate planet?

What do you do when the center goes? The upheaval brought not only a new piece of science; it brought a new cosmology. And cosmology rules. What is the big plan, and is anybody out there? Is the universe friendly? How do you think about your world when your universe has just been taken apart, its center moved? Does God know where we are now? Who are you, if you're not a little below the angels? The art world had discovered perspective, and so, apparently, had the rest of the scholarly world.

As I write this book the strength of democracy around the world has been challenged by draconian power hungry politicians during the Covid-19 pandemic. This was not the first time authoritarian personalities have exploited social and political instability for their own gain, though for many citizens it is the first time to live in such a time. How to respond, when our foundational concepts are openly challenged and basic convictions seem undermined? Where is the normal we left behind only months ago? Everything is unstable, everywhere around the world has come unhooked. Or unhinged.

Three hundred and more years after the Copernican revolution, Western people are still in chaos over how to live, still searching for answers to questions of Being. What's more, we are also facing a similar reshaping of cosmology. More on that as we go.

FIRST VIEWS – SHOCK

People who have experienced the disconcerting altered state of consciousness which is the theme of this book know something about this kind of dislocation. Whether the altered state event was blissful or agonizing, it has required reassessing, even reinventing, their reality. Even the most beautifully memorable experience can lead to major confusion and what feels like a hostile disruption of "normality." Because of my topic, I have a hard-to-shake reputation in some circles as being "negative." Is my comment here about the beautiful experiences' leading to confusion simply my negative take on a positive reality? No, not at all.

Dr. Alex Lukeman, PhD is a near-death experiencer who used to be a psychoanalyst before he became a novelist. In other words, he knows what it's like. In a single packed statement which I have quoted repeatedly, he described the effect of a powerful altered state experience. The experience is, he says,

> …the ego's encounter with the underlying unconscious and transcendent dynamics of the numinous [the Sacred], and the accompanying destruction of traditional and habitual patterns of perception and understanding, including religious belief structures and socially accepted concepts of the nature of human existence and behavior. [ii]

In other words, sometimes *everything* blows apart. He is not being negative; he is being factual, and we need to be prepared.

I almost *always* use the Lukeman quote because it is the most powerful description I know of. When the force of the underlying unconscious bursts into the open and goes face to face with Ego! When the transcendent dynamics of the holy—wow!—explodes in your face! And every bit of your patterns and habits, the way you see and understand, and what you believe about God or no-God and everything you've ever been taught about how to be in the world—poof! Shazam! Gone! Now go rebuild. But where to start?

Tobin Hart, a professor of psychology at the University of West Georgia, took a more evolutionary perspective in the *Journal of Near-Death Studies,* JNDS:

> As elaborated across a variety of traditions, destruction is a mate to creation or growth. Something must give way, must die or be cleansed in order for something new to be birthed. Old leaves must wither before new ones push themselves into life. This is the force of death, destruction—the transformative fire that burns away the dross to find the purity.[iii]

The death in a near-death experience will at least partially take place in this life, and the same dynamic holds true after a splendid spiritually transformative experience. The seed has to break before the plant can grow: At a very surface level of our brainwork, we've known this for years. But we never thought it would happen to *us!*

What is it like, this crumbling, this coming undone? Here is another quote used in *Dancing Past the Dark* and repeated here because it's that good: a reflection on the shock and challenge of seeing one's body "from an outside vantage point" during a first out-of-body experience (OBE). The shock applies equally when there is no OBE but simply the realization of Otherness.

> To…know my vision is real, is to be exposed to the unthinkable. The world cannot be as I have constructed it; it is unimaginably different. It constitutes the death of everything I have come to know and depend upon. I am not who I thought I was, and the world is not as I assembled it. I have entered a realm that is Wholly Other, and I have not the faintest idea what it is or how to negotiate it…I no longer know who I am. I have lost all certainties. Nothing is dependable. Anything can happen.[iv]

With or without a self-observing out-of-body experience, a strong altered state of consciousness rips apart one's sense of reality. Things believed impossible suddenly appear true—reunions with dead family members, meetings with previously unknown siblings, encounters with monstrous and terrifying entities, tsunamis of knowledge about subjects never dreamed of, explosions of colors never seen on earth. Disruption of those 'traditional and habitual patterns of perception and understanding' makes a huge mess. Initially, it may leave a crater, nothing.

Reiterating the Haule quote, "The world cannot be as I have constructed it; it is unimaginably different. ...I am not who I thought I was, and the world is not as I assembled it. I have entered a realm that is Wholly Other, and I have not the faintest idea what it is or how to negotiate it"

Familiar beliefs and mental structures may have been blown apart, but something has to take their place if the person is to function. It is not a tidy process. In my case, there was nothing recognizable about the entire experience, from flying over the hospital roof (not frightening at all, which on reflection was alarming because the activity itself should have been impossible) to communicating circles-with-authority in a minimalist setting in space. But...but...Creation! What about Creation? What was that non-place? Where was God? If nothing is real, what is the point, or the sense, of history? Overwhelmingly, incomprehensibly, what does it mean to be an illusion? Now that nothing is real, what is real? Oh, the world! Creation!

For anyone in the 21st century with a background in Eastern thought, or metaphysics, or serious mysticism, some eye-rolling may feel instinctive. The time, though, was 1962, in New England, with Congregational churches. In the northeast of the U.S., far from Pacific influences, neighbors were Methodists or Quakers, not Buddhists. Charles Tart's first book on states of consciousness was still seven years in the future, along with Woodstock. The Beatles had not yet been to India. Yoga was not on Main Street. It is hard, now, to believe there was so little incorporation of Eastern ways!

Like many experiencers of powerful and frightening altered state experiences, my first approach after the hospital was *not* to approach, to distance myself from engaging with the feeling level of the NDE so as to function in the everyday world. In fact, I fled. Whatever the flaws of that psychological method, it tempered despair, or at least kept it under some control. In the process, and however unwittingly, I began unconsciously assembling furnishings for a new reality. In real-world terms, I lived in the mess for a long while. Then the ideas began emerging.

ALTERED STATES OF CONSCIOUSNESS

What is consciousness? It seems so simple! To experience ordinary consciousness is to be awake and aware. Everyone knows consciousness has levels, like a dimmer switch, from high alertness down the to the blurry region where sleep begins (un-consciousness) and then deeper into that to coma and on to oblivion.

Your partner who is sleeping is not awake, but can be aware enough of temperature changes to throw off a blanket. The young girl deep in a daydream is awake but unaware, like a spaced-out driver oblivious to approaching sirens, or the student whose mind is elsewhere. Many of us have stories of a loved one who is non-awake in coma yet responds somehow to hearing a favorite song or the voice of a family member.

An *altered state of consciousness i*s a temporary change in a person's normal mental state, which may or may not include being considered unconscious. Altered states of consciousness can be created intentionally with drugs or physical practices, or they can happen by accident, associated with illness, or spontaneously.

> "If consciousness refers to the subjective awareness and experience of internal and external phenomena, states of consciousness refer to the spectrum of ways in which experience may be organized."[v]

A *transpersonal* state of consciousness relates to peak experiences, altered states of consciousness, and spiritual experiences. One of the founders of the field of transpersonal psychology, the psychiatrist Stanislav Grof, has defined *transpersonal* states of awareness:[

> "The common denominator of this otherwise rich and ramified group of phenomena is the feeling of the individual that his consciousness expanded beyond the usual ego boundaries and the limitations of time and space." [vi]

Throughout history, this type of experience has been considered religious, a term which in the Western nations today is often met by aversion; more on this topic later. The closest approximation to a neutral term is "numinous," a term coined by 20th century German theologian and philosopher Rudolph Otto:

> This state is described by Jung as imposing itself on the subject independent of his will. It is mysterious, enigmatic and impressive, defying explanation. There is confrontation with tremendous and compelling force implying meaning both fateful and attracting. One can open one's self to it but cannot conquer it. Encounter with the numinosum is seen as an aspect of all religious experience.[vii]

Jungian scholar and psychoanalyst Lionel Corbett, in *The Religious Function of the Psyche,* explains that according to Otto, the very essence of holiness is this inexpressible quality which cannot be put into thoughts or words. The numinous, Corbett says, "grips the soul with a particular affective state, which Otto describes as a feeling of the *'mysterium tremendum'*.[viii]

It is the heart of religious experience (and you may want to use the word 'spiritual' instead) and feels like objective truth from beyond the self. This would be the powerful feeling so familiar to those who have had this kind of experience, a feeling called *ineffability*, about which person after person has said, "I just cannot put it into words. There's no way to describe it! You have to have been there."

The feeling levels of the religious experience are intense, with a sense of having encountered "otherness." They include a perception of "placeness," and sometimes directionality: movement up (sky) or down (underground). And they typically center on a perception of unmistakable *presence*, of "more than human." We know from the sacred writings of many religions that these kinds of personal experiences have existed as far back as there are records of human spiritual awareness; in fact, they are the pivot points around which the religions began. They stand on ancient conceptual ground:

> *Sacred*, the power, being, or realm understood by religious persons to be at the core of existence and to have a transformative effect on their lives and destinies. Other terms, such as *holy, divine, transcendent, ultimate being (or ultimate reality), mystery, and perfection (or purity)* have been used for this domain. [ix]

No wonder these altered states have generally been considered a religious experience, or perhaps "religious-like" or philosophical, depending on one's attachment to doctrine, and "spiritual" in today's climate. They rarely include any specific religious insignia (doctrine, ritual, or religious sign), except that a spiritual presence may be named by experiencers according to their own tradition: for example a Christian may encounter what he or she perceives as Jesus, whereas a Hindu may describe a spiritual entity as Krishna or Shiva. However, because the altered state stubbornly shows no enthusiasm for advancing the doctrines of any specific religion, it is considered by fundamentalists of any religion to be a satanic influence. People tend to fear the unknown so it is natural for rigidly dogmatic people to label the unknown as somehow evil.

The modern sacred

In the almost-one-hundred years since Rudolph Otto published *The Idea of the Holy*, sea changes have swept through theology and religious practice in the West. Women are in the pulpit, gays are in some places openly in the clergy, Jews have been back in Israel for some seventy years, 'thee' and 'thou' have been retired from many Bibles, and literal readings of scripture are under question. Mostly, God has been niced up.

In re-reading *The Idea of the Holy* this past week, I was struck by two things. First, I was having difficulty hearing the reactions of today's near-death experiencers in what Otto was saying about religious experience. Nowhere in Otto did I find the kind of overflowing, unconditional love which saturates many NDEs, and which made one woman almost rapturous about her impending death. Asked what she expected, she cried, "Oh, all that love! I just want to run up to God and hug him!" Although Otto mentions love, it is restrained, not the effulgent, overwhelming love of the modern experiences.

Second, I noticed how much Otto's tone reminded me of reading my grandfather's prose; the two men were born less than a decade apart, one in Germany, the other in Wales; both religious scholars. They had wide denominational differences Lutheranism is liturgical and somewhat austere, whereas Welsh Congregationalism was non-creedal, less formal, warmed by its hymn singing tradition. What the two men shared is a quality of religiosity which is rare today. Their attitude toward God is entirely of awe and enormity, of incomparable almightiness, which shows with Otto's leaning toward a ponderous awe and "afrightment." The lovingness of their God is that of the Victorian father who loves with serious purpose. One would not "run up and hug" Jehovah.

A different tone prevails today. There is, for instance, an influential 20th century Scottish academic and pioneer in the unusual field of secular religious studies. (*Secular* religious studies! This is very new!) His name is Ninian Smart, whose writings put forward the idea of religious experience as an awareness of an "invisible world" holding the concepts which are the determinants of our actions and expressions. He says we will characterize the determinants as "God," "the laws of physics," or "the unconscious mind," depending on our culturally conditioned worldview.[x] You will notice he includes all three as *religious*.

These religious experiences, he points out, "are not some sort of superfluous tack-on or projection, rather they are grounded in the very nature of human beings. They are…a 'natural' product of humankind."[xi] What this brings is a welcome normalization of this kind of experience, a recognition of its seriousness and spirituality (religious) without the taint of weirdness inherent in the usual terminology of *supernatural* or *paranormal*.

Mystical experience

As a base of identifying components of the altered state, we can begin with the classic four characteristics of a precise state of mind described in 1902 by psychologist William James[xii] as a "mystical state of consciousness." Note that the word "God" is absent.

1. Ineffability. A mystical experience defies expression

and words cannot fully relate it to others. It must be experienced directly to be fully understood, and the mystical experience cannot be directly transferred to others.

2. A noetic quality. Although mystical states are similar to states of feeling, they also seem to those who experience them to be states of knowledge. They are experienced as states that allow direct insight into depths of truth that are unplumbed by our mere intellects.

3. Transiency. Mystical states cannot be sustained for long. Except in rare instances, half an hour, or at most an hour or two, seems to be the limit beyond which they fade into the light of common day.

4. Passivity. James writes that in mystical states of consciousness, "the mystic feels as if his own will were in abeyance, and indeed sometimes as if he were grasped and held by a superior power." Mystical experience is a form of self-transcendence, and the mystic will often say that she or he has merged with something greater... [xiii]

James does not mention what many consider a critical feature of the mystical experience: that is a sense of unity, or the experience of becoming one with all that exists.

Walter Stace added four other dimensions of mystical experience: sacredness, positive mood, paradoxicality, and transcendence of time and space. He identified an *extrovertive* mystical experience as unity at the innermost core or inner reality of all things despite their diversity, individual identity, and separateness, and an *introvertive* mystical experience as a complete dissolution of the self, and loss of all sense of boundaries and content, sometimes referred to as "the void."

Stace considered the introvertive experience more complete. (Stace, 1960a).

Quite a variety of experiences share all or most of the qualities James and Stace ascribe to mystical experiences. The more common categories of experience are described below.

Out-of-body experience

Introducing readers to a new edition of his book *Out-of-Body Experiences: How to Have Them and What to Expect,* Robert Peterson writes:

An OBE is an experience in which you seem to be consciously apart from your physical body. What you expect once you get there: such as:

- Having a ghostly "astral" body
- Floating or flying
- Passing through walls and other solid objects.
- Seeing your own physical body like another inanimate object in the room….[xiv]

For an example of reaction to a first out-of-body experience, see the descriptions in Chapter 2.

Near-death experience (NDE):

Although the term near-death experience and its acronym NDE were not coined until 1975, comparable experiences had been known from ancient times. Indeed, says the author of one study, "Near-death experiences are probably as ancient as humans. They cut across the boundaries of time, geography, culture, and religious belief."[xv]

For many years IANDS, the International Association for Near-Death Studies, was the world's only organization focusing at these experiences. This is the introduction to their extended description:

A near-death experience, or NDE, is a profound psychological event that may occur to a person close to death or who is not near death but in a situation of physical or emotional crisis. Being in a life-threatening situation does not, by itself, constitute a near-death experience. It is the pattern of perceptions, creating a recognizable overall event, that has been called "near-death experience."

Across thousands of years and in cultures around the world, people have described powerful experiences that follow this general pattern with its common features. At its broadest, the experiences involve perceptions of movement through space, of light and darkness, a landscape, presences, intense emotion, and a conviction of having a new understanding of the nature of the universe. [xvi]

Other descriptions of near-death experiences speak to the expectations of their own readership. In 1992 *The Dictionary of Modern Medicine* offered its own view:

A phenomenon of unclear nature that may occur in patients who have been clinically dead and then resuscitated; the patients report a continuity of subjective experience, remembering visitors and other hospital events despite virtually complete suppression of cortical activity; near-death experiences are considered curiosities with no valid explanation in the context of an acceptable biomedical paradigm; the trivial synonym, Lazarus complex refers to the biblical Lazarus who was raised from the dead by Jesus of Nazareth. (Segen, 1992, p. 483) [xvii]

Spiritually transformative experience (STE):

An STE is much like an NDE but without being necessarily associated specifically with death. From the American Center for the Integration of Spiritually Transformative Experiences comes this definition:

An experience is spiritually transformative when it causes people to perceive themselves and the world profoundly differently: by expanding the individual's identity, augmenting their sensitivities, and thereby altering their values, priorities and appreciation of the purpose of life. This may be triggered by surviving clinical death, or by otherwise sensing an enlarged reality.

There are many types and many names for experiences that can share common features and be catalysts for spiritual transformation. One of the best definitions I have read of these altered states comes from an Australian experiencer, writer, artist, and designer, Linda Cull. She writes:

A Spiritually Transformative Experience is a natural, and often spontaneous occurrence of heightened consciousness by which a person experiences an ultra-reality that comparably makes their ordinary, material-based reality, appear less real or illusory. They can be catalysts for a major change in the life of the experiencer, affecting their daily life, family, community and the world they live in.[xviii]

Exceptional human experience (EHE):

The term "Exceptional Human Experience" was coined by prominent American parapsychologist Rhea White, who had what we now know as an NDE during her junior year of college. She later wrote,

"Exceptional human experience is an umbrella term for anomalous experiences that transform the individual who has them so that they are engaged in a process of realizing their full human potential, which makes the experience an exceptional human one.

The Exceptional Human Experience Network has a different approach to anomalous, out-of-the-ordinary

Exceptional Experiences (EEs). By taking the emphasis off of proof, or artificially trying to "cause" or stage events in the laboratory, or passively collecting case reports, we are actively trying to understand what these types of experiences and the experiencers are telling us as a whole.

If an experience does not have any lasting effect on the experiencer, it remains simply an anomaly, and so can be viewed objectively as a one-time happening, now finished. However, some anomalous experiences become personalized. They become part of the experiencer's life. They have become exceptional experiences (EEs). These, in turn, can initiate a process that has ongoing transformative aftereffects.[xix]

Observations

There seems to be no end to the range of situations which can produce this type of altered state of consciousness, quite commonly changing people's lives. There is no single broadly accepted term for it in English, either to define the state nor to name it, so the state of consciousness is known by many names, as here (NDE, STE, EHE, OBE) At the same time, its universality is well documented.

They have been called peak experience, spiritual experience, mystical experience, altered state, and more. Many events may produce a similar altered state. A 2019 article in a journal of neuroscience included its own list of ways in which the altered state may be accessed: meditation and prayer, sensory deprivation/isolation, music listening by "deep listeners", breathwork, and ingestion of classic hallucinogens such as psilocybin.[xx]

I know from reading IANDS correspondence over the years, their provenance seems boundless; people have reported deep spiritual experiences during suicide attempts, tooth extractions, and organ transplants, during deep meditation and profound sex, while brushing their teeth, swimming, or falling out of an airplane, as well as during a kundalini opening, and spontaneously during their own birth (yes, a newborn's memory of an NDE, conveyed by artwork in adulthood). Endless options!

The prominent parapsychologist Rhea White identified 240 specific types of exceptional experiences and grouped them into seven broad classifications: death-related, desolation/nadir, encounter, healing, mystical, peak, psychical.[xxi]

The diversity of circumstances and variability of accounts have created a fair amount of confusion about what, exactly, can be considered a near-death experience. Recognizing the many ways in which these life-transforming experiences can occur should help resolve a few issues which have been contentious for years.

The term "near-death experience" was coined by Raymond Moody, MD, author of the seminal *Life After Life*, to describe the *circumstances* of the participants in his study. They had been declared clinically dead or were life-threateningly ill. Within a short time, as the book's popularity widened, other people began reporting on their similar experiences from very different situations. The term near-death experience quickly lost its attachment to the original *circumstances* (being certifiably close to death) and became glued to the reported *experiences* (even when death was not an issue).

Ten years after the book's publication, at the IANDS office at the University of Connecticut, a quick survey of the first-person NDE accounts in its files revealed that fewer than one-third of the writers said they had been declared clinically dead, while the rest described having been ill; in surgery, , childbirth, or in an accident; or having a spontaneous experience. What the two-thirds did *not* report was that they had been clinically dead.[xxii] Most of those accounts might today be more accurately be called STEs—though that term also has its significant literal drawbacks—or "NDE-like."

However, *near-death experience* was the only term the public knew, and that is the term which has stuck, accurate or not.

A significant experience might include several elements, though lives have changed with even a single one of sufficient intensity. The elements been pretty much the same across all altered state types. The contents are like a buffet table from which a diner is not expected to take one of everything: out-of-body experience, tunnel, movement through space, darkness, light, encounters with or visions of spiritual entities, encounters with or deceased family or friends, intense emotions including unconditional love and peace, intense unpleasant or horrible emotions, a landscape, swift thought processing, messages, a life review, vivid insights, massive learning, awareness of being not in control, a sense of being gripped by the experience for a reason.

Where the clinical death experiences differed most was in the reactions afterward. It is one thing to wake up knowing you have had a transcendent larger-than-life immersion in another reality, which will take work to understand. It is a much bigger deal to wake up with a similar memory and be told by a health care professional that for a little while you were considered to be dead. Those who have been declared clinically dead tend to have more psychological issues after an NDE than those who were not, though even without clinical death, some will find their experience traumatic enough to need support. Like other people who have survived a cardiac arrest, those experiencers are likely to benefit from time with a survival or stress support group.

This confusion about death and the experiences has been hard on patients, a few of whom wanted to sue their physician for malpractice on the grounds that the NDE "proved" there had been a cardiac arrest which no one had reported to the patient. For others, there seems to be a competitive edge: "I wasn't just *near*-dead; I was *dead* dead!" often with a hint that the person's experience was of a higher caliber. The intensity of thinking oneself *in* death is not a trivial thing. There is a clear tendency today, with a few participants on some social media forums, to claim that experiencers with no clinical death or who have the "wrong" features in their account ought to be expelled from the group. Fortunately, group moderators usually jump on those complaints in a hurry. There is no single "right" kind of NDE.

The confusion remains. Participation in recent medical studies of near-death experience has required a verified clinical death. Beyond the neighborhood of rigorous academic studies, real NDEs continue to occur, and the great majority of accounts continue to come with no medical assessment of clinical death and usually with no medical record at all, Further, there is no single pattern and no single element to certify the existence of a genuine experience; a tunnel is not always present in an NDE, nor a life review or message, nor encounter with presences, nor even the brilliant light which does not hurt the eyes. A genuine life-changing experience may consist of simply hearing a voice or sensing a presence.

You get what your experience gives you, and you can tell whether it is a true NDE by how it has affected you and your life. If the way you see the world has shifted and the way you live your life changed, it was a *bona fide* altered state experience—an NDE or STE or EHE or NDE-Like—or Beverly, whatever you want to call it!

WHAT THE EXPERIENCE GIVES

The obvious difficulties with distressing NDEs come *during* the experience. The sometimes life-changing pain for those who had a beautiful NDE comes *afterwards.*

Lovely experience, difficult afterwards

Especially in the early publicity about near-death experiences, reports about beautiful NDEs and ecstatic descriptions from their experiencers led to the notion that having one somehow exalted a person almost to sainthood. The public was told that many NDErs had lost all fear of death, had become unconditionally loving, were less materialistic and wanted only to be of service to others; they were more spiritual, and some had new intuitive abilities. These were true reports, and they all sounded so good and meant so much to those who were grieving loved ones, no one wanted to mention that there was a downside. A parallel truth was that family, friends, co-workers, and especially the experiencers were having a hard time.

Spirituality had shot up, but family and work relationships often hit a low. "I don't even know who she is any more!" A police officer who had treasured watching football with his son could no longer stand the violence of the game. An executive had been looking forward to a substantial promotion with a large jump in salary, and his wife had been looking at bigger houses; but now, after his NDE, he found the job didn't seem to matter: "Maybe we could just take a couple of years and go work with the poor in Asia somewhere." A hotshot accountant on a corporate fast track was back from the hospital, frustrated and feeling morally desperate because "I never noticed it before, but all they think about at work is money!" What no one had expected was that experiencers would come back alienated from their culture and former way of life.

While public attention was showering experiencers with lofty expectations, the NDErs were suffering. Almost all were homesick for "home," a few homesick to the point of considering suicide to get back to the light. Many were divorcing. Lifelong circles of friends broke up because an experiencer no longer fit. Church choirs lost formerly reliable soloists, who lost the joy of group singing because their beliefs had changed.

One of the early IANDS groups in Connecticut was meeting regularly with a mixed group of experiencers and interested laypeople. Conversations at the early meetings were invariably lively, centering on hearing someone's experience account, with questions and discussion afterwards. The experiencers were happy to share at first, and the non-experiencers were thrilled and inspired to hear first-hand encounters; they said it meant so much, just to hear the accounts. After a while, though, the leadership noticed that the NDErs were becoming quiet. A bit more time and the non-experiencers were complaining that they were the only ones participating in discussions. No one knew quite what to do.

Then came a visiting speaker, a well-known experiencer from out of state who had done a good deal of television. There was a good crowd that night, as many people had come to hear her experience. To their dismay, she talked hardly at all about her NDE. What she was talking about was the difficulty she was having adjusting to life in her new identity, how lonely she felt among old friends, the difficulties with her kids' reactions, the painful disconnect with her husband. As she talked, the NDErs began to join in, and very quickly, for the first time in a long while, they were all contributing, adding details from their own struggles. Suddenly one of the observers jumped to her feet in agitation, wringing her hands and obviously distraught: "Oh, please, you mustn't talk like that! Oh, that is not right! You have been to the light—tell us what we are to do!" And the experiencers stopped talking and were silent again.

The early news of near-death experience, so filled with light and love, had lit such profound hope in an aching and anxious world, the idea that it might come with difficulties seemed like a cosmic betrayal. If unconditional love did not wipe out all suffering, how could there be any balm for a contemporary Gilead?

That is, in fact, a major thread in this book. How do we respond when what we thought might be salvation disappoints? How are we to understand these things? There may be a long way to go and a lot of very hard work to find peaceful integration. There is even more work in leading non-experiencers to understand that a glorious NDE is not a magic button.

Distressing within NDE and afterwards

Even for people who do not have a clear religious perspective on hell, our culture is so saturated with symbols left over from medieval and earlier times, everyone has to deal with some of the overflow. The fiery lakes and torments of Dante's Hell are very much a part of contemporary mythology.

I have been sad to discover how many people's first assumption about a distressing NDE is based on an immediate sensation of falling, or the image of a goat, or a snake, with the assumption that they are headed for a guilt-based external hell with all its Dantesque implications. Almost as many slam the door on any mention of religious belief and call the images hallucinations (and then discover that the images sit just on the other side of that door and do not go away on their own.).

Whatever their personal specific religious beliefs and despite space travel and Hubble photographs, residents in Western societies have absorbed the three-tiered cosmology which says God/the Sacred above, Earth in the middle, Hell somewhere below our feet with all its gory imagery. Despite my liberally spiritualized religion, I have always lived in this Western culture and had to struggle through some of those same doctrinal swamps.

First, there is no tested research evidence that having an unpleasant NDE means the person is evil, mean, angry, unloving, God-hating, depressed, or criminal. Those adjectives will apply just as well to people who have blissful NDEs. Having the NDE does not mean you have been assigned to hell after death. Simply not true! We do not know exactly how it works, but some people who had awful NDEs later have had heavenly ones, and that includes genuine saints. Some people with multiple NDEs have both types.

All NDEs are vivid, and some are, without question, horrifying: violent, or ugly, or menacing, or loathsome; some include terrifying monsters. In some, people feel themselves being torn apart, or eaten…or eating; or feel the events as meaningless despair or as the presence of ultimate evil. Images seem magnified, huge and distorted. Look at what the writers say about them on page 18; the images, the feelings, the forces are *enormous*. Even the hyperbole in which they are described is so exaggerated as to be meaningless: "A million times worse…," "ten million times more horrible"… "three billion times more painful!" The feelings are agonizingly real, but the essential fact is that the experience is spiritual, not physical; it is not literally a factual event but a powerfully experiential encounter. The body is not bleeding or in pieces when the person wakes up. While the message will be in some sense true, the experience must not be understood as literal fact.

Hell. "Am I going to hell?" I dealt with the hell question at some length in *Dancing Past the Dark*, and have put that chapter on the website, *www.dancingpastthedark.com*. Several reputable authors have written on the subject; a bit of reading about the books on Amazon will tell whether the book is a balanced, factual approach to the subject of hell or an evangelical sermon.

The short answer about the conclusion of that study is that despite what literalist readers of scripture believe, the Bible does not contain the hell of Augustine and Anselm and Dante. Simply not there. The view of eternal physical torment developed hundreds of years after the Bible was written. It is possible to be logical about this *factual* discovery and to disconnect from much of the sickness and terror by cognitive work. That's the easy part!

What is far more difficult is to find release from the grip of the archetypal images of the imaginal psyche. It is one thing to disregard a mistaken teaching, but *what was that experience?* Unfortunately, most religious institutions say little if anything helpful.

Hell or not, *What was that?* The answer is part of my story of integration. Read on.

Identity. Another question treads on the hem of the first question's garment. Whether or not I am going to hell, that altered state experience is still present, and it is clearly not affirming. Other people have beautiful experiences, and some of them say they have done a lot worse things than I ever did. *Something* must be very wrong about me. What did I *do?* Is my DNA bad? What is there about me that is so bad that...that *thing*...is what I get? This is a much bigger question, and with a longer answer. It is, for the most part, the question it has taken me almost sixty years to resolve. If this book does not hold some actual answers, I hope it will at least point in some interesting directions.

Religious belief. I was deeply confused after my NDE. That place had not been on earth, which meant it had to be in God's territory. But that experience had not shown so much as a quiver of godly presence, nor any sign of occupation beyond those circles. Maybe the angels and Jesus and his disciples were elsewhere, but there should have been *something?* Maybe it really was about hell, but neither had there been a quiver of judgment or punishment. It was all spiritual bureaucracy.

Where could one even go to get information? Question after question, and no answers.

---

[i] Tarnas, Richard. *Cosmos and Psyche*. New York: Plume, 2006.

[ii] Lukeman, Alex. Book review of Edward F. Edinger, *Encounter with the Self*. Retrieved from *Tiger's Nest Review*. Created 3/25/98; accessed 9/4/01. http://www.frii.com/~tignest/encount.htm

[iii] Hart, Tobin. "Integrating Spiritual Experience: Peaks to Plateaus," *Journal of Near-Death Studies,* 33(2) Winter, 2014. 86-99.

[iv] Haule, John Ryan. *Perils of the Soul: Ancient Wisdom and the New Age*. York Beach, Maine: Weiser, 63.

[v] Https://www.academia.edu/4246524/Altered_ States_of_ Consciousness _and_Transpersonal_Psychology

[vi] Grof, Stanislav. *Realms of the Human Unconscious: Observations from LSD Research*. New York: Viking, London: Souvenir Press. 1975, 1993.

[vii] Http://Encyclopedia.Uia.Org/En/Development/12338.

[viii] Corbett, Lionel. *The Religious Function of the Psyche*. London: Routledge, 1996, 11.

[ix] Https://www.Britannica.Com/Topic/Sacred

[x] Http://People.Bu.Edu/Wildman/Relexp/Reviews/Review _Smart01.htm

[xi] Http://www4.Westminster.Edu/Staff/Brennie/Rennie% 2019 99%20 CSSR.28.3.63-69.Pdf

[xii] Http://Web.Csulb.Edu/~Plowentr/William%20 JAMES.Html

[xiii] Https://www.Ncbi.Nlm.Nih.Gov/Pmc/Articles/ Pmc6707356/

[xiv] Peterson, Robert, Charles Tart. *Out-of-Body Experiences: How to Have Them and What to Expect.* Newburyport, MA: Hampton Roads Publishing, 2013.

[xv] Masumian, Farnaz. "World Religions and Near-Death Experiences," in Holden, J., Ed., *The Handbook of Near-Death Experiences.* Santa Barbara, CA: Praeger Publishers, 2009, 159-184.

[xvi] International Association for Near-Death Studies. https://iands.org/ndes/about-ndes/what-is-an-nde.html.

[xvii] Segen, J. C. (Ed.). (1992). The Dictionary Of Modern Medicine. Carnforth, England: Parthenon.

[xviii] Https://Lindacull.Com/2015/11/Whats-a-Spiritually-Transformative- Experience/Religious.

[xix] White, Rhea.

[xx] Barrett F.S., Griffiths R.R. (2017) Classic Hallucinogens and Mystical Experiences: Phenomenology and Neural Correlates. In: Halberstadt A.L., Vollenweider F.X., Nichols D.E. (Eds) Behavioral Neurobiology of Psychedelic Drugs. Current in Behavioral Neurosciences, Vol 36. Springer, Berlin, Heidelberg.

[xxi] Https://www.Encyclopedia.Com/Philosophy-and-Religion/Other- Religious-Beliefs-and-General-Terms/Miscellaneous-Religion-43

[xxii] A hand count of 'clinical death' mentions in accounts in the IANDS office files, circa 1985. The finding was never written up. Ken Ring's files were separate and may have included a larger percentage of clinical death records.

CHAPTER 3.

RITES OF PASSAGE

There was no clue, during my immersion in anthropology and multiplicity of perspectives, how important that background would become. Five years later came the NDE, and although the effect was not immediate, that saturation in shamanism and ethnography silently began to do its work. I believe that what saved me from the typical grip of the fear of Western hell was that I knew about tribal rites of passage. I had read the work of the great Romanian/American shamanic scholar and historian of religion, Mircea Eliade!

Just in the span of my lifetime so many traditions which were taken for granted have morphed into a new normal. For example, when I was a child, Memorial Day was special; yet today is Memorial Day in the United States, and it comes at an awkward time. We are in The Year The World Shut Down, the year of covid-19, and across the country large gatherings remain, if not forbidden, certainly frowned upon; the traditional Memorial Day parades have been canceled, along with the once-mandatory civic remembrances of our dead, the Brownie Scouts and Cubs placing flowers on military graves, the generations of local politicians lined up on shaky platforms one more time.

In those many years when Memorial Day was sacred time, at an agreed-upon hour, the entire nation stopped for a full minute of silence to honor our dead. From coast to coast there was a hush as radios went silent, business stopped, mothers kept their hands still, and conversations ceased. Together, we all remembered. The parades which followed were legendary; I remember being awed into silence by the sight of two Civil War veterans in the back seat of a convertible in Rochester, New York, each of them holding a little American flag.

Today it is mostly considered a beach day. With more emphasis on Casual Friday, praise bands, rebellion against Political Correctness (which used to be simple courtesy—ah, my age is showing), we are unused to having to sit through process for its own sake. Mostly this is, I think, a good thing; but some traditions cannot be un-formalized without significant loss. There are no speeches, tedious or otherwise, at town parks this year, and no young ones will be able to remember seeing veterans of World War II nod from wheelchairs.

"Ritual events known as rites of passage represent another important example of social sanctioned institutions that have provided experiential training in dying in ancient civilizations and ancient cultures."[i] That is Stanislav Grof talking, one of the giants of transpersonal psychology.

That stops us. What? *Experiential* training in…dying?"

We are so naïve. In the industrialized 21st century West we tend to think of rituals as associated with happy social occasions—marriage, graduation, maybe a retirement party. First a quick little ceremony, usually at church, then on for food and music and dancing. Party time! After all, we came out today for a parade. But Mircea Eliade, the great scholar of shamanism, throws a more serious tone to the party: *Pg x:*

"Initiation is equivalent to a basic change in existential condition; the novice emerges from his ordeal endowed with a totally different being from that which he possessed before his initiation; he has become *another.*"[ii]

Rites have always been associated with religious, or at least reverential, occasions for a very good reason: they point to an active intersection between the sacred and the local people. (The sacred may be God, or a god, an unknown sacred force, a power-holder; but notice, in traditional terms there is no quibbling about the existence of a greater power, a sacred *something.*) Religion has been a primary vehicle for art, music, song, dance, and other forms of aesthetic experience.

Important rites, then, are related to sacred involvement in major understandings, and in life changes which have social significance for the entire community. Their literal meaning stretches back sometimes through millennia, even when recognition of that meaning has eroded. Traditionally, major rites lead to celebration, but they have always included pain.

The major Western initiation into adulthood is getting a driver's license. The traumatic elements of indigenous rituals horrify Western sensibilities: scarring, stabbing, impaling, terrifying, life-threatening and violent acts. For some deeply symbolic rites, blood must be shed (circumcision, blood from the sexual parts associated with new life, life is in the blood). Even today, an initiation for young children marks them socially as belonging and carries sexual implications for the future—the *bris,* circumcision, for eight-day-old Jewish baby boys, and, in a good many sub-Saharan cultures, circumcision of six-year-old girls.

The rites are never considered to be punishment, even though some extreme types of initiation around puberty will produce agonizing physical pain, cognitive dissonance, and even PTSD. Traditional societies recognize that extreme pain can lead to an ecstasy which is understood as direct communication with the deity. Hallucinogenic drugs are sometimes involved, particularly when the objective is to initiate a shaman, whose experience is likely to be filled with demons, monsters, extreme violence and ritual (spiritual) dismemberment. From the community perspective, pain and challenge are what any initiation is about—a test of the ability to come through, and a doorway to hidden routes to the deepest realms of consciousness. We find them barbaric; but we are missing some depths. Our pain is more typically boredom.

Of these intensely disruptive initiations, Eliade says, "It is primarily these ordeals that constitute the religious experience of initiation—the encounter with the sacred. The majority of initiatory ordeals more or less clearly imply a ritual death followed by resurrection or a new birth…Initiatory death signifies the end at once of childhood, of ignorance, and of the profane condition."[iii]

All kinds of near-death and spiritual experiencers can certainly be considered a spontaneous rite of passage, although the social situation is far different. Whereas a traditional rite is expected, planned for, understood and supported by the community, for us these altered states of consciousness come out of the blue, with no informed helpers in attendance. Until very recently, most Western people had never heard of such a thing. Even when an experience appears as a dim vestige in religious literature, the lines are so blurred by time and shifting understandings, there is rarely any detail about the difficulties involved. All we see is that Jesus went into the desert and the Devil came and argued with him; then angels came and took care of him, and he moved to a different town. The Buddha sat under his bodhi tree; and then he got famous. We do not see their inner struggle.

What Eliade describes is not the kind of experiencing of the sacred today's language of spirituality talks about; yet is a kind of spirituality which seems always to have existed and which is still reported. Somehow, we must come to terms with that as historical and psychological fact without throwing either the experiencer or the sacred out of our culture altogether. I do not go in search of pain; but I do believe we must find ways to acknowledge its power. We must stop being so afraid of the not-nice, even of the horrifying experiences. They have things to teach us.

At a private seminar, eighteen NDE experiencers gathered for a long weekend to talk about how their lives had been affected. With one exception, their NDEs had been wonderful; yet every person had come to the weekend hoping for help with intense struggling. The seminar leaders later observed, "Participants described the NDE as an unanticipated, dramatic, and complete immersion into a reality unlike anything experienced in their lives previously. [They] underwent what some described as a permanent and complete paradigm shift in their reality and view of themselves, a sudden correction in their accustomed path and perspective on life… These new outlooks, directions, and changes created friction in almost all important areas of the participants' lives. As one woman commented, 'I had to put the pieces of my life back together in a new way.'"[iv]

Trauma comes with many faces. The blatantly disturbing scenarios of distressing altered states torment from *within* the experience, whereas transcendent and otherwise beautiful NDEs throw up what can be unspeakably painful challenges *after* the return home. Eliade explains about the life-passage "It is impossible to participate in a new irruption of sanctity into the world or into history, except by dying to profane, unenlightened existence and being reborn to a new, regenerated life.[v]

In other words, there is good reason behind the Fundamentalist requirement that its members must be "born again," though we need not accept the rest of their theology. With metaphor as bridge, we may be able to get where we want to go. The only way for an organism to have a new life is that it must lose the former one, which condition demands an experience strong enough to do the job, powerful enough to grab our attention and not let go until we respond. As a culture, we have not been helpful; too many of us are offended by the notion that the spiritual could be other than lovely and inspiring. Time to get serious. Go deeper.

EXPERIENTIAL PHASES

The pattern of rites of passage is universal, and has been for thousands of years. Whether bland or tortuous, any rite of passage has three phases. These phases operate whether the rite is minor, such as a school awards ceremony, or major, a state funeral, and whether they are intentional, as baptism or bris, or spontaneous, as with Howard Storm's NDE.

Separation

In the first phase, neophytes withdraw, or are withdrawn, from their current social level in preparation of moving to another status. In cultures like ours, which take the rites relatively lightly, the preparation may be as simple as dressing the baby in ceremonial clothes before Baptism; confirmation classes before church membership; the haircuts of basic training to "cut away" the civilian self.

Transition

The transition phase moves from cognitive information to direct experiences of altered consciousness. The shifting is introduced, as Stan Grof has noted, by "technologies of the sacred" such as drumming, loud music, chanting, rhythmic dancing, and probably some sacred psychedelic drug. There will be sometimes excruciating pain among the challenges, as an ability to deal with suffering is a measure of one's readiness. In *Sacred Pain,* author Ariel Glucklich notes, "A child becomes an adult, or an outsider becomes an insider when ritually controlled pain weakens the subject's sense of empirical identity and strengthens his or her sense of attachment to a highly valued new center of identification."[vi]

Grof continues: "Such experiences typically take the form of psychospiritual death and rebirth and encounter with numinous dimensions of reality. In the context of such rituals, these experiences are then interpreted as dying to the old role and born into the new one."[vii]

It is, I think, not possible to convey in words what the visions in these rites would have *felt* like to the novitiate. The images in these tormenting events, which appear in myths and poetic sagas, would be in-your-face close, breathing on you—monsters with terrifying teeth; demons; weird insectoids; the pregnant belly of a mysteriously beautiful woman; nameless creatures tearing your flesh, eating your heart; spiritual beings watching silently; deadly beings from the deep sea, reaching for you with poisoned arms, picking your bones clean. ... He who has been successful in such an exploit no longer fears death; he has conquered a kind of bodily immortality... [The images] emphasize not only the danger of the passage ...the paradoxical nature of passage into the beyond, or more precisely, of transfer from this world to a world that is transcendent."[viii]

Incorporation

In the third phase, the initiates re-enter society with a completely new existential status. "Newly initiated persons are not the same as those who entered the initiation process. They have undergone a deep psychospiritual transformation, they have a direct connection with the numinous...they have faced and survived a convincing experience of personal annihilation, they have transcended their identification with body and ego, lost fear of death, and have attained a new attitude toward life.[ix] Does this sound familiar?

SPIRITUAL PHASES

Whatever the type of event—tribal rite, NDE, STE, mystical breakthrough or other—the spiritual responses parallel the physical events and follow a generally similar pattern. Eliade said much the same thing:

As some psychologists delight in repeating, the unconscious is religious. We can still recognize the patterns of initiation—in the spiritual crises, the solitude and despair through which every human being must pass in order to attain to a responsible, genuine, and creative life...[We become ourselves] only after having solved a series of desperately difficult and even dangerous situations; after having undergone "tortures" and "death," followed by an awakening to another life, qualitatively different because regenerated.

The pattern remains although our vocabulary changes to reflect the cognitive perspective: a Christian view will likely understand the phases as representing *suffering, death, and resurrection*, while a secular outlook will call them *conflict, crisis, and resolution,* and a metaphysical vocabulary could be *suffering, end of a mode of being, and open to spirit*.

Writing in the *Journal of Near-Death Studies,* psychology professor Tobin Hart shared a presentation he made at a conference of ACISTE, the American Council on the Integration of Spiritually Transformative Experiences. Three archetypal forces, he says, are typically awakened in this kind of altered state:[x]

Expansion

The spiritual event is commonly experienced as providing an *expansion in awareness* of some sort. It may occur dramatically, as a kind of explosion into another perspective, or instead as a more gentle coming to awareness. In either case, the common factor is that it provides some opening of view or possibility—a first-hand immersion into or taste of a previously cloaked world. A synonym for this expansion of awareness is transcendence.

Destruction

Destruction may be experienced literally as a health crisis or a near-death experience, but it is also apparent psycho-spiritually and metaphorically through any STE as one's old worldview collapses in the face of new experience or as one's ego, values, and self-definition wobble or even crumble.

Communion

Communion emerges as the third force commonly at work. An STE is sometimes paradoxically experienced as both new and deeply familiar—a profound homecoming, perhaps, or sense of belonging. In a unitive or oceanic experience there is a sense of fusion or communion in which the experiencer no longer perceives separation between oneself and the world. Sometimes there is a sense of being held, guided, or supported. Experiencers sometimes interpret this communion as an invitation back to one's soul's path or as recognition of some cosmic benevolence.

In traditional initiation rituals, the purpose was not only to create Eliade's "basic change in existential condition" understood as a sign of maturity and a source of pride ("You did it! Good job!") but it was also expected that the ordeal might give access to a glimpse of other realms than the ordinary world. In some cultures, obtaining such a vision was (and is still) an essential aspect of the trial, a gift with which to return to community life.

And here again is our weakness. What do we see in the confusion of returning to a reality which, at least after a beautiful experience, is such a letdown after the serenity of the NDE? John Ryan Haule calls it "the return to the broom-closet." Too often, a medical establishment which thinks NDEs are "curiosities with no valid explanation" sends the experiencer home with a prescription but no supportive word. Culturally we have very little available to help the experiencer and family understand what has happened and how to deal with it. The person struggling with the social conflicts of a post-radiant NDE needs more than "get over it."

And surely we can do better with distressing NDEs than to see them as punishment for who-knows-what—proof of spiritual incompetence, a sign of moral failure, and lacking in the approved positive thinking!

The West, Eliade points out, "has played the central role in bringing about a subtly growing and seemingly inexorable crisis—one of multidimensional complexity, affecting all aspects of life…To say that our global civilization is becoming dysfunctional scarcely conveys the gravity of the situation." (p. 11)\ As hospital chaplain David Maginley commented, "Here we have ego deconstruction, once again. How foreign this concept of ritual initiation is to the western ear, tuned to some fantasy of self-actualization and independence: 'I am my own person!' We are a people who have lost our ritual vocabulary."[xi]

Wherever a rite (or an altered state of consciousness) occurs, it will be governed by such cultural assumptions. And when crisis comes, culture will determine what kind of response to offer and how to deal with it. It is easy to understand that joy, with or without angels, is more appealing than existential terror. But as accounts throughout history demonstrate, darkness is also a natural and recurring aspect of human experience. If we do not explore it intelligently, we go uninformed and unprotected into those territories. Why our extreme resistance, the stubborn refusal to look at distressing NDEs? Where we resist most strongly, that is where we need to be looking.

---

[i] Grof, Stanislas. *The Ultimate Journey: Consciousness and the Mystery of Death.* Ben Lomond, CA: MAPS, 2006, 35

[ii] Eliade, *Rites and Symbols of Initiation: The Mysteries of Birth and Rebirth* New York: Harper and Row, 1958, p. x.

[iii] Referenced in David Stephen Calone, *The Spiritual Imagination Of The Beats,* Cambridge University Press, 2017. 91.

[iv] Stout, Yolaine *et al.* "Six Major Challenges Faced by Near-Death Experiencers," *Journal of Near-Death Studies 25:1, Fall 2006, 49-62.*

[v] Eliade, Mircea. *Rites and Symbols of Initiation: The Mysteries of Birth and Rebirth.* New York: Harpercollins, 1980, 118.

[vi] Glucklich, Ariel. *Sacred Pain: Hurting the Body for the Sake of the Soul.* p. 7

[vii] Grof, *Ultimate Journey,* 38

[viii] Eliade, *Rites and Symbols,* 64-65

[ix] *Ibid.*

[x] Hart, 88.

[xi] David Maginley, personal correspondence

PART II.

WHAT WE DISCOVER

CHAPTER 4.

THE 'PARA' REALMS

And then I went to work at IANDS. Everything from this chapter on I have learned since that pivotal job application. It is like the moment in the movie *The Wizard of Oz*, when the screen goes from black and white to technicolor. One way and another, it has all been about integration and growth and Big Ideas.

At no point until I was working at IANDS did I ever approach the situation of my experience thoughtfully and analytically, because I had no concepts, no language in which to think about it. The 'predestination' answer essentially delivered itself in the hospital almost before I had even asked what that experience was all about, and I rejected it without having a replacement. Where 'that dream' was concerned, my needs were, first, to keep the unexplained experience away from my daily life and years later, after the divorce, to keep the lights on and three kids fed, so the "It" would have to look out for itself.

Beyond that, my psyche was working on its own, and it went inside and talked to itself. The integration of my NDE has been a great unfolding punctuated by inexplicable synchronicities (which word I did not know when this all began).

Since humankind came out of its infancy, it has seemed clearly evident in the mainstream world that our universe is dualistic: body and spirit. However, only in the last few hundred years has anyone believed that our physical senses and spiritual energies occupy non-communicating worlds. They used to talk to each other; yet since the Western mind discovered rationalism and industrial technology, the physical and spiritual realities have been curiously like locked cars on a train, traveling together yet going to different places.

Science in our time has taken on the status that formerly belonged to religion, with its own fundamentalisms. One tenet of that fundamentalism is that only those things which are measurable can be demonstrably real. This creates some awkwardness for spiritual activities, which have not stopped happening but which, by not being physically measurable, are considered by Scientism to be "fake news." In scientific terms, because they cannot be reliably counted and measured, they are considered 'beside or beyond the normal': *paranormal*.

'SOMETHING IS GOING ON.'

Twenty-five years after my anthropological immersion, that work experience was about to become more practical than I could ever have guessed. All that practice in multiplicity and perspectives!

It does not take long, in the country of altered states of consciousness, to discover that, from the perspective of scientism, one is not in Kansas any more. When the first of those file drawers opened, out tumbled letters describing conversations with dead family members (many of those letters); a question about how a four-year-old returned from the hospital talking about a sister who had been miscarried years before his birth (he said he had met her during his surgery); testimonials to being unafraid of death; one person's lengthy description of correspondences between several of the writer's past lives and the present one; and a question asking who in the area was good at dowsing so the writer could have a new well dug in the right spot. In sum, the letters were coming from a conceptual space just across the border with what would be considered the common wisdom in 20th century America. More letters were coming every day, and phone calls.

Before six months were out, there was the phone call from an apprehensive woman in another state saying there were lights moving in a circle over the head of her bed, and could I tell her how to get rid of them? (I could not, but one of the Board members could, so I relayed the message, and the woman called back later to say thank you, it worked.) A letter arrived from a stranger halfway across the country, answering a question I was asking in a letter not yet mailed. There was the deeply intuitive experiencer who simply showed up at the office one day to say she had received a message from her guides to come and offer her services. She became my first non-student assistant and one of my closest friends. (I honor your name, dear Leslee Morabito. RIP.)

There were so many out-of-the-mainstream stories, it was not altogether surprising that at home, when we needed to have the septic system cleaned but could not find the access, my visiting older daughter said, "Maybe you could ask someone at the office? Someone will know where it is."

My new context was an almost total overthrow of mid-twentieth-century mainstream "normal." I had never seen fairies or communicated with spirits, had not consciously known anyone intuitive (psychic), and there were no family stories to pave the way. I had never before known anyone who had come back from the dead, or who talked with disincarnates in the dining room.

It all became so much, so fast, that one day I simply left the office early and went home to sit quietly and "feel normal." The house was in a clearing on a 40-acre tree farm, with a little garden plot where I had recently put in sugar snap peas. This day I took a cup of coffee out to the deck for that peace and quiet, but when I looked out at the garden I gasped and tipped over the coffee.

Little green tendrils were poking up where that morning there had been only bare dirt. My heart was pounding as if in a panic attack, and I was overcome because *they were doing this all by themselves!* In the context of all the reality-shifting happenings at the office, this was simply a head-snapping smack of pure truth: There is more to the world than what we can see and measure. Something that isn't physical is doing things. *Something is going on!*

Of course it sounds absurd; after all, I had planted those peas. What did I expect? I had had Victory Gardens as a kid; of course I knew peas come up. But this is what enlightenment is about, and how it begins: I had never before *seen* and *felt* the mystery in what was happening. All at once, everything became alive. I had begun to pay attention to what science ignored.

Firmly in the grip of fundamentalist materialism for the past three centuries, Scientism was no help in understanding these phenomena, labeling everything I was encountering either misguided, pseudoscience, paranormal, or psychotic; their condescension was thick enough to spread.

On the other hand, I had already encountered a number of people whose metaphysics kept them mentally drifting without ever quite touching the ground, and a good many experiencers who had only their own literal comprehension (or incomprehension) of their NDEs to go on, yet who broadcast it as dogma. I was seeing too many experiencers whose unrestrained egos were in a condition of grandiosity and social dysfunction or bleak disarray. ("I wouldn't expect you to understand," said one of the star NDE experiencers to a volunteer. "My consciousness is higher than yours.")

Where were the level-headed voices with genuine information on which to build new understandings? Finding them would take time.

BENT SPOONS

'If we can do this, what else can we do?'

The occasion was a meeting of the IANDS Board of Directors, early in the 1980s. Eight men—they were all men—came into town for a long weekend three times a year, or maybe four, treasuring the time to nurture the embers of this fledgling, one-of-a-kind organization, three days of real conversation with others who did not look at them oddly when the subject of "all this" came up. It was so early…there was no money…they were doing this on their own dime (including paying the sole staff member's salary), meeting in private homes to avoid hotel costs and staying with local IANDS members or sleeping on the host's couches. By today's standards, Board meetings were, in a word, funky.

This meeting was in the home of one of the lead researchers, a century-old mill, tall, thin, historical house, a brook almost touching the rear wall and a small state road not much more than ten feet outside the front door. It was nicknamed "The Near-Death Hotel" because so many visiting experiencers had stayed there.

The directors were an interesting group: the earliest three NDE researchers (after Moody, who was never an active presence), all academics, two of them physicians and one a psychologist; one attorney; another academic psychologist; a Pentagon specialist in the paranormal, whose wife was traveling with him; a health care administrator; one wealthy person with deep interest in a high level of "weird stuff." Professionally, the researchers all walked tightwires in academic departments where science—a deeply skeptical scientism—was the rule; they had all suffered discipline or demotion for their persisting interest in the taboo "funny stuff" subject of near-death experiences. The only women present were the Pentagon man's wife and the IANDS office manager, newly titled "executive director." The host's partner had made preparations for the meeting, but she was out of town for the weekend.

Despite the seeming edginess of their interest in NDEs, some of the Board members were still new to the paranormal "all this," still cautious about what it was safe or possible to believe. So when, late in the morning's discussion, the Pentagon man mentioned that he and his wife had recently learned to bend spoons using only their minds—great interest! A high-energy conversation ensued about the how's and whether's, with opinions ranging from "Are you serious!" to "old news."

I don't remember whether the decision was by acclamation or volunteerism, but at some point it was agreed that the Pentagon man's wife would undertake a research demonstration. A small group hurried to the kitchen to select the spoon. It must be metal, something sturdy; the committee presented several options which were rejected as having weak design features or feeling flimsy. Finally, the host dug around in a different cupboard and came up with a winner: his partner's best serving spoon, almost a foot long, high quality stainless steel with no indentations on the handle and a round, not flat, neck. No flimsiness here!

The spoon was, indeed, so sturdy that the researching spoon-bender looked dubious.

"I don't know. It's awfully big…but, here, let's give it a try."

She settled back in her chair and set to work, stroking the length of the handle, rubbing the neck gently, getting the feel of the spoon, looking more and more thoughtful as she worked. Several minutes passed with no change. Then more minutes. The Board went back to their agenda. A half hour or so, and it was time for lunch.

"I don't know," she said. "This is taking a long time. It may not work."

Her hands were still busy with the spoon, which she kept stroking as the group made space on the long table for lunch. Sandwiches and more conversation. A few people walked outside to look at the brook. The meeting was about to resume.

And then, a shout: "I can feel it!" and those at the table could see her hands make a sure but not Herculean bending motion, and the heavy handle was bent double, and the spoon's bowl almost touched the handle.

Bedlam! Witnesses had been with the Pentagon man's wife the entire time. There had been no tools, no tricks, no hanky-panky. We could see what she was doing, which was simply, thoughtfully creating friction on the metal. And thinking. As she said,

"Willing it to bend."

She said she had moved energy down her arm to her hands, and from there to the spoon. Everyone had to hold the spoon, to marvel, to express their shock or satisfaction. Vindication! Everyone wanted to hear about what it felt like, what the spoon-bender had been thinking during the long process. Pictures were taken. Norma's spoon was the central attraction of the weekend. Oh. Norma. Norma would want her best serving spoon back.

As I drove away late that afternoon, my rearview mirror showed a group of men on their knees on the asphalt road, trying with tools and physical force to bend the spoon back before Norma's return.

This is how scientific open-mindedness advances, sometimes one spoon at a time. No one who was at that meeting would ever again wonder if spoon-bending could really happen. One researcher later said he spent his entire flight home bending the airline's plastic spoons.

It was my formal introduction to "all this."

How does this place really work? If we can do this, what else can we do?

QUANTUM MECHANICS: THE LOSS OF SUBSTANCE

The pre-Copernicans thought they were cosmically stable; we grew up thinking we were physically solid.

It was old news: Two and three thousand years ago, in what would become the West, everyone knew the earth was like a snow globe—Heaven and the deity far above, Earth a flat disk in the center, the underworld below, all of it surrounded by water. It had been created for us, all the plants and animals, sea creatures and mysteries, and we were to care for it and each other. For at least 2500 years we believed that with our whole hearts as a culture, give or take a few doubters, and after Copernicus we still mostly believed in the Creation part and taking care of Earth and each other. At least, we said we believed it.

As the Enlightenment progressed, science stayed over here and religion went over there, and they stopped talking to each other. Science made the universe work like a great machine, and then evolution came with lots more questions and doubts about God's role, but those of us who were more into faith than physics had a hard time understanding what science was saying, because it didn't fit religiously with what people believed—or were trying to believe—and what the culture was still teaching.

And then, early in the 20th century, along came quantum physics, and pretty soon the social split widened, because on the physics side it was just too much crazy talk that made no sense, and besides, who could understand all that mathematics!

Seven years before I was born, a British astronomer and physicist stood before an audience at the University of Edinburgh to deliver the year's Gifford Lectures. He began by telling them the story of his two tables.[i]

> I have settled down to the task of writing these lectures and have drawn up my chairs to my two tables. Two tables! Yes; there are duplicates of every object about me – two tables, two chairs, two pens.
>
> This is not a very profound beginning to a course which ought to reach transcendent levels of scientific philosophy. But we cannot touch bedrock immediately; we must scratch a bit at the surface of things first. And whenever I begin to scratch the first thing I strike is my two tables.
>
> One of them has been familiar to me from earliest years. It is a commonplace object of that environment which I call the world. How shall I describe it? It has extension; it is comparatively permanent; it is coloured; above all it is substantial. By substantial I do not merely mean that it does not collapse when I lean upon it; I mean that it is constituted of "substance" and by that word I am trying to convey to you some conception of its intrinsic nature. It is a thing; not like space, which is a mere negation; nor like time, which is – Heaven knows what! But that will not help you to my meaning because it is the distinctive characteristic of a "thing" to have this substantiality, and I do not think substantiality can be described better than by saying that it is the kind of nature exemplified by an ordinary table
>
> My scientific table is mostly emptiness. Sparsely scattered in that emptiness are numerous electric charges rushing about with great speed; but their combined bulk amounts to less than a billionth of the bulk of the table itself. Notwithstanding its strange construction it turns out to be an entirely efficient table.

It supports my writing paper as satisfactorily as Table No. 1; for when I lay the paper on it the little electric particles with their headlong speed keep on hitting the underside, so that the paper is maintained in shuttlecock fashion at a nearly steady level. If I lean upon this table I shall not go through; or, to be strictly accurate, the chance of my scientific elbow going through my scientific table is so excessively small that it can be neglected in practical life.

The external world of physics has thus become a world of shadows. In removing our illusions we have removed the substance, for indeed we have seen that substance is one of the greatest of our illusions. The shadow of my elbow rests on the shadow table as the shadow ink flows over the shadow paper. It is all symbolic, and as a symbol the physicist leaves it.

Do you see what he just did there, this amiable astrophysicist, knocking down our foundations? This is our Copernicus moment, the time, for us, when it all comes unglued. The kicking-apart of trust and stability, understanding, and what makes everything worthwhile, and our families, and the Creation, an *illusion?* How are we to understand Being in this immateriality?

Decades later, my NDE would tell me the same thing: "You are not real. This is all an illusion." I have not believed that any more than most physicists have believed it.

Illusion is a commonplace in Eastern thought, but here in the West we are grounded (literally) in some four thousand or more years of a philosophy and religion which prize their place in a tangible Creation, an intentional and beautiful physical place of *things*—soil and body and blood and growing things and relationship. There is no history of *illusion* here other than the wiles of the Evil Ones who would tempt us into unbelief. Sir Arthur Eddington had a hard road.

His science was so new, most of his university audience knew nothing about it. The ordinary world of ordinary people certainly knew nothing about it. Only a minority of students get as far as taking physics in science, and that is largely Newtonian. In sixteen years of schooling, I heard nothing about Eddington's table. And thirty-four years after his speech, when I had an NDE that said neither the Creation nor I was real, I did not know about his table. One of the earliest things IANDS did for me was to put Eddington's table and the work of a whole school of physicists directly in front of me, to learn on my own. It was painful, and terrible, and it had to be done.

Sir James Jeans wrote:

> 'The stream of knowledge is heading toward a non-mechanical reality; the universe begins to look more like a great thought than like a great machine. Mind no longer appears to be an accidental intruder into the realm of matter, we ought rather hail it as the creator and governor of the realm of matter. Get over it, and accept the inarguable conclusion. The universe is immaterial—mental and spiritual.'[ii]

An article in the online publication *Collective Evolution* reports the Dalai Lama's claim that spirituality, without quantum physics, is an incomplete picture of reality. He says, "At the center of every atom is only energy."

> If you observed the composition of an atom with a microscope, you would see a small, invisible tornado-like vortex, with a number of infinitely small energy vortices called quarks and photons. These are what make up the structure of the atom. If you focused in closer on the structure of the atom, you would see nothing, a literal void. The atom has no physical structure, thus we have no physical structure, and physical things really don't have any physical structure! Atoms are made out of invisible energy, not tangible matter. ... Quantum mechanics is essentially the science of consciousness and spirituality, proving just how connected we are to everything in existence, which is all an illusion. [iii]

Buddhists have an easier time of it with quantum theory; at least that's my conviction. They grow up grasping illusion as friendly territory. Even the brilliant Western physicists who were reporting all these conclusions have had their own difficulties. Theoretical physicist Richard Feynman told an audience:

> What I am going to tell you about is what we teach our physics students in the third or fourth year of graduate school... It is my task to convince you not to turn away because you don't understand it. You see, my physics students don't understand it, that is because I don't understand it."[iv]

The physicists were saying that nothing is materially real, that neither trees and cows or we are what we think, that from the perspective of our atoms we are buzzy little electrical fields made up almost entirely of space. Quantum mechanics says that we're built on indeterminacy, that nothing holds still, and that what is invisible may be more substantial (at least metaphorically) than what we can see and count and measure, that science can itself fall down rabbit holes. But most of us had not heard that, though it has been subtly reshaping our world for a century. Schools didn't teach it unless they were universities. Only very smart people knew what was going on, and none of them could explain it. In the predominantly Christian West, the common folk might as well have been still living pre-Copernican lives, except we knew about the sun.

A CONFUSION OF ADVERSARIES

Right about in here, it seems to me, I hit an impasse because of so many conflicting voices. If the atom has no substance, this is a scientific fact; how is it a religious or philosophical issue? The answer, I think, is going to be something related to their once being together but then separated by the Enlightenment;

but at this point in the discussion there are not yet enough pieces to create a new argument.

If the atom is at base not a thing but a spark, then so are we and so is everything—*every thing*—and its solidity, its stability, its knowability vanishes; the universe may very well be an intelligence, but that has no tangibility; we may be making it all up.

For some of us it takes time to get past the sense of betrayal at the destruction of the Creation we had not realized was our rock. The anger—like street protesters after yet another death of a black man at the hands of the mostly white police—the anger is made up mostly of our grief. This is what happens when we first encounter our ontology, our Being, and are told it does not exist.

To those who don't get the anger or the sense of divine abandonment inherent in this new physics, there may be less feeling involved. The anger of a good many people now, in the twenty-first century, is directed not toward physics but toward religion itself, or at least what has been made of religion.

To those of you who have known Western religion only as rules and rigid coldness, impositions from an attenuated bureaucracy with no sense of your reality, knowing it as exclusion, as threats and repression and accusations, I am—we are—very sorry. That's not what it is supposed to be like. You have seen a hard shellac, not the good wood beneath it. The substance. Its spiritual intent is to tell the story of a power which makes things whole, which operates out of relationships and yearnings to bring healing and justice wherever they are in danger. The heart of the teaching is responsibility and love. In its maturity it looks to embrace all people, to tend the fruitful Earth, and to learn justice.

To be told, "There's only you," and even then, you are illusory—this becomes the ultimate, meaningless aloneness.

How can one love when there is no one *to* love. Hollowness. We have to think our way to a new way of everything.

BACK TO EDDINGTON: 'IF SCIENCE IS TRUE, WE HAVE NO RECOURSE.'

How can a truth of no-thingness coexist with an embodied cosmology? Late in his Introduction, Eddington addresses the tension between mathematics and ordinary communication:

> It is difficult to school ourselves to treat the physical world as purely symbolic. We are always relapsing and mixing with the symbols incongruous conceptions taken from the world of consciousness. Untaught by long experience we stretch a hand to grasp the shadow, instead of accepting its shadowy nature. Indeed, unless we confine ourselves altogether to mathematical symbolism it is hard to avoid dressing our symbols in deceitful clothing.

As has been abundantly noted, I do not speak mathematics; as a language, it eludes me almost completely. But the truth is unquestionable, and what I must do is sit with it long enough for the terror to subside.

> Every cell in the body obeys the principles of physics…but the fact that physics speaks in mathematics removes it from the experience of living.[v]

The trick is to borrow the findings of mathematicians and astronomers and move them into words, while attempting never to overstate nor to sentimentalize their symbolic meaning. We live after all, in the same universe, sharing the same reality—whatever that may mean—while attempting to comprehend each other's language.

Take-away: If science has demonstrated the utility of quantum mechanics—as it has—and has shown as much as can be spoken mathematically about its truth, then we have no recourse but to listen and talk about what we have heard. And to learn to do so without PTSD. Inch by inch…

---

[i] Eddington Http://Mathshistory.St-Andrews.Ac.Uk/Extras/ Eddington Gifford/.
[ii] Henry, R.C. "The Mental Universe," *Nature* 436:29, 2005.
[iii] Https://www.Collective-Evolution.Com/2017/04/26/Dalai-Lama-Spirituality-Without-Quantum-Physics-Is-An-Incomplete-Picture-Of-Reality.
[iv] https://mathshistory.st-andrews.ac.uk/Biographies/Feynman /quotations.
[v] Chopra, Deepak. *Infinite Potential.* Foreword.

CHAPTER 5.

STAGES: THE UNFOLDING CONTEXT

The integration of a transpersonal experience is as much a process as is the growing of a human being from infancy to adulthood; only mechanical devices are amenable to linear assembly. Human process is unique, complex, highly individual, and not given to straight lines. The route is often an elliptical path cycling some steps forward and then a few back before once again inching or leaping forward.

In this section I have decided to summarize a therapeutic process and three separate stage theories because each of them bears on altered state experiences in a special way. It was in placing them together that I became aware of their structural parallelisms and the implications taking them beyond simply mechanical importance. We start with four psychological/behavioral stages of integrating post-traumatic stress; then common developmental stages in spiritual and religious growth from infancy to adulthood; four mutations in the history of consciousness; and close with five stages in the history of a belief system.

With regard to NDEs, Cherie Sutherland[i] in Australia described a variety of "integration trajectories" of testing and finding social acceptance, a process lasting from the experience itself until it becomes a part of the experiencer's life. She found blocked trajectories in clients whose experience was considered meaningless, either by the individual or by those to whom it was disclosed. Not knowing what to make of an NDE could lead to an arrested trajectory, as could fear of disclosing or absence of social support.

In my instance, my integration would have been considered both blocked and arrested, as the experience made absolutely no sense to me and I had not the foggiest notion what to make of it, except as noted throughout this text, that all meaning and purpose had been wiped out of existence if Being is illusion, and/or that I had encountered (and rejected) predestination. I was in no mood for nihilism. A rough outline of my stages appears in the post-traumatic stress section below.

Visionary thinker Ken Wilber describes stages as "progressive and permanent milestones along the evolutionary path of your own unfolding."[ii] Each individual person and experience is unique and yet every living thing, both individual and social, plants as well as creatures, moves through a series of phases, or stages, toward a mature version of its biology and consciousness. Each stage has its own characteristic behavioral and conceptual abilities, its own strengths and challenges, and goes beyond its predecessors yet includes and integrates them in its growth. Something of our past is always present with us as the ground of new growth. At each stage some things are carried forward while others are lost.

1. POST TRAUMATIC STRESS THERAPY STAGES

The literature and practice of post-traumatic stress therapy can be helpful as an entry point to dealing with a traumatic NDE or other altered state. Although this most obviously applies as a response following a distinctly hellish experience, it is also helpful as a means of dealing with any difficult re-entry. I wish we talked about it more. It is not new; Bruce Greyson, one of IANDS' founders and the preeminent academic researcher of near-death studies, was studying PTSD and NDEs back in 2001.

"Trauma is an emotional response to a terrible event." That's the American Psychological Association (APA) definition.[iii] The event is in the past, over and done with; it cannot be erased, but the emotional *response* must be dealt with. Integration is the process by which one can transform terrible to livable and better, in spite of the memory and effects of the trauma. It takes place over time, in stages, and it is not guaranteed because many people don't want to do the work.

Here, for example, are four stages defined by a PTSD clinic. I did not have PTSD (or any other) therapy, but, in preparing this section, easily recognized both the stages and how I had responded to them. It has to be noted that although most people work through their integration in a reasonable amount of time, mine was almost interminable.

PTSD stage 1. Tight focus on the trauma.– immediate recoil and literal interpretation of remembered traumatic images; psychic numbing. If not addressed, this can go on for a long time. Needs time, patience.

NEB: *"No!'* Total recoil. I could not speak about the experience to anyone for 20 years. Six years after the NDE, when I saw a picture of the circle, it was labeled Yin/Yang, which I understood only as terror. I threw the book across the room, bolted out of my friend's house, and ran home, locking the door behind me. *"No!"*

PTSD stage 2. The near view. Able to back away from the trauma a bit. 'What does this mean for me?'. Grief, anger, isolation. Personal ideological/religious adjustments. Breakdown. Begin reframing,– If there were another way to see this, to understand this...what might it be?

NEB: A *very* long stage for me. Confrontation with the Yin/Yang begins this period, overlapping the 'No!.' I had no concepts about the NDE except predestination, which I refused to accept but which had a big role in my inability to talk to anyone about the experience. Otherwise, the whole thing was unexplainable. There was a lot of isolation, with overwhelming anger at God, at church. I had no idea what to do. Year 14, I came up with a classic strategy: I would do a Master's degree in Clinical Psychology. So the coursework began. Year 16, my visible life fell apart, beginning with a traumatic divorce. Shorthand version of the next four years: 2/3 empty nest, career gone, house gone, Master's unfinished, civic position gone, no answer about the NDE. Not only was I not real, but all my labels were being torn off, and I had no money for therapy. The next three years were spent hitting bottom. The fires had to burn themselves out. By the end I had a new and supportive partner and I was reading voraciously: especially the mystics, Joseph Campbell, Alan Watts, John Sanford, Morton Kelsey, the developmental psychologists.

PTSD stage 3. Seeing the bigger picture.
Longer term theoretical personal issues, a major work period: shadow work, revisioning.

NEB: For me the PTSD stage 3 theme should be in lights. It had taken a long time, but 20 years after the NDE I was hired to run the office of a tiny startup nonprofit, the International Association for Near-Death Studies, IANDS. My experience got a name, though it was not welcome at the office in the heyday of heavenly NDEs. But I learned about experiences other than mine, and discovered desensitizing and began to be able to approach my NDE, and to talk objectively about others; it was possible to keep a folder of distressing NDEs which arrived in the mail. Ten years into IANDS, Bruce Greyson and I wrote the first research-based paper about distressing NDEs, published in the journal *Psychiatry*. I became able to tell my story publicly, though still unable to articulate a meaning or purpose for it. I was still reading 24/7, the NDE research literature, Carl Jung, Stanislav Grof, Ken Wilber, Charles Tart, Caroline Myss (who became a friend).

PTSD stage 4: Possible integration: May be more shadow work, emphasis on nature of being, compassion, service to others. The arm swings back to the universal mode of trauma recovery: compassion and service—but they are the mark of trauma, not the mark of NDEs only; they are post-traumatic *growth*. Suffering moves us to compassion and service.

NEB: Another 20 years: Nobody said these stages would all be brief! Fifty years after my NDE, I was able to write and publish a comprehensive, research-based overview of distressing NDEs, *Dancing Past the Dark* (which was referred to in a recent IANDS keynote address as "the bible of these experiences.") I felt my public mission was completed. Well, not quite completed.

~

Many people want simply to get rid of the trauma, to get it to disappear. The problem with that is that it will go underground, but it never goes away and will erupt on its own terms. Integration is more likely when we acknowledge the confusion and pain and work through it, rather than around it. To reach that point, we need to ask more than 'What was your experience?' We have to get inside its meaning for the owner. The stories are vital, but we need more than just their skin. In that sense, summarizing my NDE above was useful but superficial, merely description. We really need to ask, what was transformative? What opened doors?

The process: Grief, rage, depression. Facing the demons and dragons. Reframing. A ton of shadow work. Reading. Revising. Letting go. Finding ideas.

Now, pages of words later, we come to the first of the revelations which now shape my understanding of that long-ago-yet-always-with-me NDE.

2. PERSONAL DEVELOPMENT THEORY

The 20th century gave us more than the wars to end all wars. Among its most productive gifts was the emergence of psychology as a discipline, most especially in the work of developmental psychologists.

In Switzerland, Jean Piaget, in the 1920s, was the first to recognize, through meticulous clinical observation and working with his own children, that every child moves through four stages in their cognitive development: 1. awareness of self as a physical entity; 2. interacting with the environment both actively (magic) and symbols (art, language); 3. increasing ability with concepts, classifying, early logic; 4. development of logical thought, extending into adulthood.

Essentially, what Piaget was discovering was that biological/neurological skills develop in predictable sequences, and children use the earlier capacities as groundwork for more advanced skills. The stages are natural, sequential phases of growth and cannot be skipped. His findings have been replicated, at least in outline, in countless other studies.

I first encountered stage development theory while I was still teaching. Jean Piaget's work on cognitive development in childhood was immediately captivating because the theory answered so many dilemmas about behaviors: "Oh, *that's* what's going on!" That has been a common reaction to every development theory. Erik Erikson (1958) looked at psychosocial development and found a similar pattern;

Lawrence Kohlberg at moral development (1958/1981); developmental psychology was finding its investigators, and they were all finding similar patterns in child development.

Stage development of faith (trust)

James Fowler's work on the developmental stages of faith shows the phases in the development of faith, Faith, in his terms, does not necessarily involve religion but whatever is the primary object of trust. For some, this will be God (our usual sense of faith) but others will hold money, sex, an organization, a sports team, a natioinal identification at the core of their trust. Fowler was not a psychologist but was Professor of Theology and Human Development at Emory University. He found seven stages, the first not numbered because it is pre-personal.

A week-old infant is obviously a person, so how can its consciousness be considered pre-personal? Theoretical thinking has no trouble with this. The infant is pre-personal because it has not yet developed the ego which will think of itself as "I," as "me." The baby has only sensory awareness, not recognizing a difference between itself and the answers to its needs. It centers on itself and the mother or caregiver as being the entire universe. However, it is soaking up information about whether its environment is safe, dangerous, trustworthy, hostile. At some point in the first or second year, the baby recognizes that it is not exactly the same as the caregiver; this is the pivotal development of ego. The *Me!* of the 'terrible twos'.

Faith Stage 1.

Between the ages of two and perhaps six, although the mother-figure is still central, other people have been added to the baby's world. Language is introducing the child to its culture with its do's and don'ts and no end of stories. This is a wildly imaginative time, full of magical thinking when a wish is as good as a reality. Morality is crude, with no principle but "If you do that, you'll be sorry." The child's thinking is egocentric; the only point of view is theirs. And the developing ego needs plenty of exercise, the beginning of "No!" and stubborn resistance to doing what someone else wants done.

Faith Stage 2.

By the end of first grade, there has been a swing to masculine influence. The child's world now extends beyond the family. Authority rests with whoever is bigger and more powerful. Morality is reward/punishment. The child can imagine someone else's point of view and can begin to think in a logical way but not abstractly; examples will have to be concrete, as understanding is literal. Stories from history and myth (in other words, patriotic and religious stories) are understood as factual. A symbol's meaning is not in what it represents but in itself; loving the country and loving the piece of cloth that is a flag are equivalent, as defacing a religious painting means damaging God. This stage can last through adulthood for many people.

Faith Stage 3.

From middle school age to adulthood, authority and the child's sense of identity rest with the group; conformity rules: This is how we live, and this is what is true; other people (strangers) are "their kind, not like us." The child begins to think hypothetically and can begin to understand symbols as representative not literal. Jordan Peterson has commented, "The group is the historical structure that humanity has erected between the individual and the terrible unknown." [iv] Many adults remain here for life; it is the single largest stage.

Faith Stage 4.

Late adolescence typically brings an uncomfortable time of questions, doubt, and disillusionment, with feelings of betrayal when earlier beliefs turn out to have flaws or to be untrue. Teachers may be considered liars and former political heroes seem like frauds. Authority figures come under deep suspicion about their motives. Nothing is quite the way the child has thought. Conspiracy theories may seem to provide a sense of control. This tends to be a law-and-order stage to attempt finding less chaos and more security. Ironically, says M. Scott Peck about religious beliefs, people at the previous conformity stage usually think people at this questioning stage have become 'backsliders,' when they have actually moved forward.[v]

Faith Stage 5.

From early mid-life or later, an adult will be able to let go of the single point of view and reconcile multiple perspectives. Authority rests in the self rather than in the group. It becomes possible to recognize the limits of logic and to accept paradoxes. There is often a discovery of "who I really am," and the beginning of seeing life as a mystery. Sacred stories and symbols can be appreciated for their spiritual/ philosophical meanings rather than as literal readings. These are years of pulling things together by more broadly including others.

Faith Stage 6.

This stage is rare as a lived life, though many people aspire to it. This is the stage of a God-grounded life, inclusive of everyone and living to love and serve others. These are adults who are truly "in the world but not of it," for they pay attention to their own convictions rather than to what everyone else is doing. They are typically considered weird or dangerous by those in earlier stages; martyrdom is not unusual. This is not a matter of their perfection but of their integrity and spirit. Greatness of commitment and vision often coexists with great blind spots and limitations. (Fowler's list of examples includes Lincoln, Gandhi, Mother Teresa, MLK Jr.)[vi]

What's the point of knowing about stages relative to NDEs and STEs? Transcendent experiences often illustrate morality at higher levels, such as inclusiveness, unconditional love, universality. At those stages there is an absence of emphasis on strict "thou shalts" and mandated beliefs, which is confusing to experiencers who come from earlier stages; the difference creates turmoil when they return to homes and work environments which no longer "fit." This discordance is how accusations of demonic activity arise, because comprehension operates only as much as a person's level allows; higher level principles cannot be understood by those who are not there yet. This conflict operates all the time in everyday life with topics such as politics and religion.

Knowing where we are helps us check in with a sharper perspective. If something about our experience or our reaction is rubbing us the wrong way, perhaps it has to do with the stage we are at, or the beginning of a stage ahead that we didn't see coming.

Every stage has its strengths and limitations. The stages do not represent value judgments, though people at upper stages see the limitations of earlier ones while those at early stages may be suspicious of those ahead of them.

Stages are fluid and cyclical: some aspects of early stages remain with us as the later stages develop, and we cycle back through various parts of them. No stage is the only "right" one; each is appropriate for those who are in it.

Movement from one stage to another involves discomfort, like getting out of a cocoon, and rarely happens all at once;

we may operate out of bits from several stages simultaneously. Shift into and out of the questioning stage is especially multi-faceted, often difficult, because it is the one most centered on doubt and challenging authority. An individual moving into that stage may seem to be changing cultural or religious beliefs when what is changing is really the perspective of a different stage.

3. MUTATIONS IN CONSCIOUSNESS HISTORY

In the 1930s, a young German cultural philosopher found himself driven by a project he had in mind. His name was Jean Gebser, and his project, which had come to him as an epiphany, was to trace the history of human consciousness and describe an impending new epoch which was already beginning, an age of what he called integral consciousness. When he began, he was in Munich, but then the Nazi Brownshirts arrived, making life difficult for scholars. He fled to Italy, but soon moved on, penniless, to Spain, where he stayed for six years. He quickly became fluent in Spanish, worked in civil service, and became friends with Federico Garcia Lorca and other poets, translating their work into German. The Spanish Civil War caught up with him, and he fled Madrid only twelve hours before his apartment was bombed. He made his way to Paris, where many intellectuals were finding refuge. In the next two years he would befriend both Pablo Picasso and Andre Malraux and other notables until World War II exploded. Seeing the futility of war and another advance of the Brownshirts, Gebser escaped from Paris and, barely two hours before the borders closed for the duration, crossed into neutral Switzerland. He would live just outside of Bern for the rest of his life.

Gebser is more influential today than when he was alive. As a biography on the Jean Gebser indicates,

In a nutshell, what Gebser succeeded in demonstrating through painstaking documentation and analysis was this: Hidden beneath the apparent chaos of our times is an emergent new order. The disappearance of the pre-Einsteinian world-view. with its creator-god and clockwork universe as well as its naive faith in progress. is more than a mere breakdown. It is also a new beginning. In fact, long before the apostles of a "new age" arrived on the scene, Jean Gebser spoke of our period as one of the great turning points in human history. What makes his work so appealing and relevant is that it offers a unique perspective on human history and the present global crisis. When Gebser's study on the unfolding of human consciousness was first published it was considered one of the most controversial intellectual creations of our era. This is still true; his ideas challenge not only those of the establishment but also many of the new contenders.[vii]

When he arrived in Switzerland, it was 1939, and he had been working on his project for a decade. Another ten years of work would be required before his major work was ready. The book was published in 1949, titled *The Ever-Present Origin*. As a whole, the project is an early glimpse into post-modern thinking, as Gebser disdains the modern conventions of hierarchy, linearity, and authority; if published today, the book would be multi-media. He also deplores the notion of progress, which distances consciousness from what he sees as the centrality of Origin. *"The structures cannot be pinned down into historical events,"* he says. Gebser speaks of these structures as "spiritual processes, first and foremost, and before anything else. To merely historicize them would be a mere perspectival take."7-9 [1]

"In the rolling thunder of the immanent present, all that we *are,* all that we *have been,* and all that we *could be* is radically with us. Time is whole, and therefore *you* are whole."[viii]

The research that went into the project was monumental, covering the entire sweep of human history with detailed consideration of anthropology, cultural history, art and language, architecture, the sciences, as well as technological development and more. Most unfortunately, Gebser had severe asthma and was unable to teach. Until seven years before his death at the age of sixty-two, he was almost completely ignored by the academic establishment. It was then that the venerable University of Salzburg, in Austria, created a special professorial chair for him in comparative culturology. As his website says, "This unique appointment was a belated acknowledgement of his genius. But it changed little, if anything, in Gebser's lifestyle; he had lived and worked most of his life as a maverick." [ix] Nevertheless, he was a major influence on Ken Wilber, who took the title of his Integral Theory from Gebser's work, and on other philosophers and futurists.

For our purposes, there is no need to spend time with the intricacy of Gebser's thinking, which is astonishing, or the complexity of his overall material, which is stunning. Although his entire theory is complex, fortunately, his history of the worlds of consciousness is relatively simple to describe.

From his massive research, Gebser discerned human consciousness as having presented in five nonhierarchical structures, or epochs, with a new one beginning to unfold. He does not say they "developed"; rather, they are leaps or "mutations" which *emerge* rather than coming as linear progress. He ways:

The structuring which we have uncovered seems to provide a clue to the foundations of consciousness. This structuring rests upon the recognition that clearly distinguishable worlds have come to the fore during the development of Western society, whose unfolding took place in mutations of consciousness. We use the method of pointing out the structures of consciousness during the various 'epochs,' on the basis of their peculiar modes of expression in images as well as in languages, as revealed in valid records. [x]

He is almost aggressive in his insistence that under no circumstances should the form of development he describes be considered "progress." He does not call them "stages." Consciousness, he states, "is neither knowledge nor conscience but must be understood for the time being in the broadest sense as wakeful presence. Second, this presence or being present excludes as a contradiction any kind of future-oriented finality... We do not share...the conviction that the contemporary stage, or any rational, perspectival structure represents the *non plus ultra* of human development."[xi] But, in light of our coming to his work directly from Piaget and Fowler, just look at the epochs his study has revealed, and their order!

Unless otherwise indicated, the notes below are drawn loosely from Gebser's own descriptions, with or without sentences.

Archaic structure

Perhaps 200,000 years ago and more. What would it have been like, he wondered, to be among the first conscious humans? "It is akin, if not identical, to the original state of biblical paradise: a time where the soul is yet dormant, a time of complete non-differentiation of man and the universe." All future mutations are co-present and latent; inherently unstable, as the structures of consciousness are in some sense leaning into their own potentiation, waking to spring forth. Zero dimension: there is no sense of space or time; nothing is happening. All dominated by physiology, by instincts, by simple perception, sensations, and emotions. Nothing is they-themselves; everything is world.[xii]

Ken Wilber calls this phase "Dawn Man," whose existence is "embedded in physical nature (pleroma) and dominated by animal-reptilian impulses (uroboros), *immersed* in subconscious realms of nature and body, of vegetable and animal, initially feeling indistinguishable from the world that had already evolved to that point. This is "the structure lying behind the universal myths of a Garden of Eden, a time before the 'fall' into separation and knowledge and reflection, a time of innocence."[xiii]

Magic structure

The distinguishing characteristic of the magic structure was the emergent awareness of nature (as distinct from self). Sometime in the unknown distant past, there began a leap from zero dimension to the one dimension of pointed unity: the emerging differentiation of self from world. The word group 'make,' 'mechanism,,; 'machine,' and 'might,' all share a common Indo-European root *mag(h)-*. Gebser conjectures that the word 'magic,' a Greek borrowing of Persian origin, shares the root. In this structure lie the countless forms by which magic man tries to cope with nature. The part represents the whole; so, a drawing of a spear killing an animal is predictive of the successful hunt. Capacity to merge between magic and nature, consciousness and trance a continuum with the preceding epoch, and reciprocal movement. Humans begin to *have* a world rather than identify *as* the world, but with no taking, claiming, or conquering. With the magical structure comes the emergence of art and musical instruments and primitive sculpture. and making the magic itself. The art is *mouthless:* Only when myth appears does the mouth, to utter it, also appear. It indicates to what extent significance was placed on what was heard, not on what was spoken.[xiv] Ritual acts are inseparable from their outcomes. Material reality is symbolic reality, full of the wonder and terror of other agencies and powers; the magical is neither purely symbolic nor literal but both at once. Shamanism is comfortable here.

Mythical structure

The essential characteristic of the mythical structure is the emergent awareness of souls. Souls and stars, astronomical time and archetypal time are synonymous with the emergence of sophisticated calendrical systems (Maya, Aztec, Egypt). Two-dimensional: duality, polarity, complementarity. It is a final move out of the magical structure as nature becomes mastered through cosmology and ancient sciences. Move away from the mouthless face; the mouth must now speak. The inward-facing mythical structure is a coming to consciousness of the person-as-soul. Timelessness leaps into rhythmicity, the mythical structure expressing itself in the circularity of winter and summer, the astronomical cycles of cosmogony, the complementarity of heaven and the underworld. The polarities, timelessness and time, the eternal return and rhythmic periodicity, mundane and sacred time are complementary, not yet As with magical structure, everything, including the least consequential detail, had significance.

Myths of this structure begin to express the move from mythical structure to the mental structure through the motif of the sea voyage – humanity's precarious relationship to the deep waters of psyche. Sea as image of soul upon which go the daring heroes on adventures of self-making, pitting selves against cosmic forces and dangerous powers of the soul. Another prototypical element of a nascent mental structure: Wrath or anger bursts the confines of community and clan to spur the hero on toward further individuation and ego-emergence. Gilgamesh and Enkidu set out to battle the old nature gods of the magical structure; the result is not triumph but the existential contemplation of mortality. Another duality of the mythic: the ego is both delighted in self-becoming and agonized by the awareness of its tragic impermanence (ultimately impotence). Oceanic thinking circles around the numinous content of the world. The emerging mental ruptures the circle with directional thinking. The emergence of consciousness which effects its presence in these myths reminds us that not just the sun but also the darkness within man is thereby rendered visible.

Mental structure

The Mental structure announces the coming of the ego, the work of spatialized consciousness. Consciousness moves into linear time, into progression, the concept of historical time. Clock-like cosmos of rational order and churning of mechanical gears – specialization of time. If the Magic is the point, and the Mythic is the polarized circle, the Mental is the triangle. A momentous rupture from the mythic membrane. This process…bursts man's protective psychic circle and congruity with the psychic-naturalistic-cosmic-temporal world. The ring is broken, and man steps out of the two-dimensional surface into space. Fundamentally alters the world. This is where our enchantment ends, at least for this duration.

Space is without content, so the new individual goes forth in a frenzy of creative mutations in perspectival art, expansion of empiricism, map-making, exploration, but erodes the qualitative dimension of the cosmos in favor of a purely measurable and void matter. (feminine 'mother' exchanged for the de-souled 'matter'). With the Greeks enters 'taking apart' thinking. Ambiguity becomes impossible; must be either-or. In contrast to mythical *oceanic thinking* we have *pyramidal thinking*. Thesis, antithesis, synthesis – the triangle, Trinity, an *oppositional cosmology, heaven and a hell that must be vanquished.*

A future-oriented form of time, a movement in writing from left (the unknown) to right (known, wakefulness). Mental structure brings paternal judgment in monotheism and a genealogical line of male prophets/kings. Alphabetic culture in Greece divides spoken language into script; makes abstraction possible, moves from *image* to the *word*, from the mythos to the logos. Clarification of thinking supplants the dreamlike mythical structure's psychical and imagistic consciousness. Jesus walking on water is a dynamic expression of a new mental consciousness no longer threatened by re-immersion into the imagistic and watery depths of soul. The organ of mental structure is the eye, the insulated fortress of the modern ego. The previous reality of permeability and magical sympathy with the spirit world, or myth of eternal recurrence with the soul and World Soul, recede behind the sectored 'point of view'. Secularity. Self-discovery, revelation.

Trinity – creative, speculative, dogmatic (unity always has a magical character). Trinity renders visible 'man thinking'(speculative) or what is to come (creative). The effector, the bearer of consciousness, is the ego. Fully in the mental structure, anthropocentric structure where consciousness becomes centered. This ego proceeds from the sea voyage, the integration of the psyche, in Odysseus. The mythological bearer of consciousness was Helios, (as in the attribution of Roman emperors, *sol* Invictus), later Christ; and it was Christ who was the actual bearer of consciousness and thus empowered to lead the soul.

Christ's immunity to re-submersion into the psyche is symbolically expressed by his survival of a shipwreck; and he was able to walk on water. With this deed he overcomes the depths of chaos and is entitled to say, 'I am the light of the world.' It is at this point where the paths of mankind, East and West, are to diverge. (In India, one designation of Vishnu is Naravayana or, literally, 'He who walks on water.') Christ's deed was to a great extent divested of its magic character because of his historicity. Of more significance is its obvious consciousness-intensifying aspect.

The identical deed that prompts Christ to accept suffering via his conscious ego, leads, in Buddhism, to the negation of suffering and to the dissolution of the ego, which, transformed, returns to the original state of immaterial Nirvana. In Buddhism the suspension of sorrow and the Ego is held in esteem; and this suspension of sorrow and suffering is realized by turning away from the world. For Christianity, the goal is to accept the ego, and the acceptance of sorrow and suffering is to be achieved by loving the world. Thus the perilous and difficult path along which the West must proceed is here prefigured, a course which it is following through untold hardship and misery.

Gebser Summary

The deep, dreamless, and latent sleep of the archaic, the dreaming timelessness of the magical nexus, the imagistic power and starry rhythmicity of the soul in the mythic, and the wakeful, spatial exuberance and glorious self-discovery of the mental. Each structure expresses its own world, its own ontological reality, which we can now hopefully appreciate without assuming the wakeful, mental claim to exclusive objectivity. If the mental cannot 'measure' the measureless, it is not real, but we know that this attitude is the equivalent of an amnesiac who has been cut off from their living past. The archaic, magical, and mythical worlds live on in us if only as latent presences and sometimes ghosts.

What Gebser has seen is that the pattern of our individual comings into consciousness, as seen with the developmental psychologists, is a miniature of the great coming into being of all human consciousness. Both the infant Consciousness and the infant human move their wakeful presence through these same stages: Eden, magic, myth, mental, the underlying pattern structuring the whole.

4. BELIEF SYSTEM DEVELOPMENT

Pattern, not content

The same pattern appears in the emergence of consciousness in our spiritual institutions.

As mentioned with Active Imagination, experiencers are often troubled when the content of an NDE or STE does not match what the belief system of their former life taught them to expect. Wrestling to reconcile the beliefs can be painful and may, without help, lead nowhere. Insights offered by a developmental view are often a key to the puzzle.

The power of understanding this developmental pattern is that it enables those who are struggling with issues of belief to discover that changing ideas does not have to mean being lost or doomed but merely moving to a different place on the road. For many experiencers, this has been a life-saving revelation. Human psychology is human psychology at any scale; the phases are the same, though the language differs.

The tables below demonstrate an example of stage development in the history of religion, especially the Abrahamic religions of the West. In this case, the tables show stages of Christian ways of thinking, which have morphed into denominations, which often reflect different stages of the same religion. Stages 1 and 2 can be clearly seen in Old Testament texts and the most literal interpretations, with #3 evident in much of the Second Temple era and continuing in mainstream traditions today. The psychological development is what determines stages, whether of a person or an entire community.

Never mind whether you believe or disagree with any of the text on the charts; this is not an attempt to proselytize! The point is to see the pattern of changing ideas as the overall system moves from one stage to the next. The same exercise could be done with any religion or ideology. These charts track Christian development because they are the charts I have, and because this is the religion I know well enough to be certain of their relative accuracy.

Each of the clusters below represents a stage in the development of the overall religion. The belief topics–the *content*–of the group's faith are in boldface, consistent throughout all the stages. followed by the *interpretation* of the topics which is common at each stage. Notice that the actual content (l) of the institution does not change, while the stage-bound interpretations (right column) shift and create entirely different outcomes.

Tribal

Earliest stage known, recognizable in pre-history. (Fowler stages 1-2, Gebser Magical). Key words: Magical, concrete, literal.
Bible. Since roughly 1000 BCE. God's actual words, literally true without error
God/gods. Awe-inspiring superbeing in the sky; capricious, dangerous
Jesus Since 1st century of Common Era. Divine, sent to be a sacrifice to God's wrath.
Prayer. Ritualistic speaking to the divinity for confession, thanks, praise, request.
Sin/salvation. Our actions relative to religious teachings (Ten Commandments). Salvation is rescue from hell.
Heaven/Hell. Heaven is eternal safety in the sky land. Hell (only since rougjhly 1st century BCE), physical torment after death.

Kingdom of Heaven. After death. "This way is the only way to Heaven."
Mystical. Unquestioned and all-pervasive. Magic, full of fantasy and fear (real demons, battling angels, Satan).

Warrior

Stone Age; evident 10,000 years ago. (Faith stage 2, Gebser Mythical). Key words: Coercive, concrete, literal.
Bible. God's actual words, literally true scientifically and historically. Shows a wrathful God.
God/gods. All-powerful, a warrior god (avenging warrior of justice) Lives in the sky, exclusively male.
Jesus Makes war on sin, death, Satan. Believe in him or be doomed.
Prayer. Rituals of spiritual warfare shape world events. Tangible God, Satan, demons, angels.
Sin/salvation. Sin is our acts against God. The violent atoning death of Jesus saves us from our sins. ll.
Heaven/Hell. Heaven is only for believers in Jesus. Hell is God's revenge on sinners & unbelievers.
Kingdom of Heaven. After death, in the sky. "We have the only way to Heaven."
Mystical. Like Tribal, full of magic and fear. The power of symbols is in the physical object.

Traditional

Iron Age 5,000 yrs. (Fowler Stage 3, Gebser Mythical). Key words: Faithful, conformist, exclusive, community.
40-55% of world today; Evangelical/conservative mainline churches.
Bible. Range of interpretations; most believe Bible stories reveal truth about God, wrath.
God/gods. Righteous, punishes evil, rewards obedience. Only

right way. Male, outside of creation.
Jesus God in human form, a miracle worker ; atoned on the cross for our sins.
Prayer. Ask God up there to do things for us down here, but also thanks and praise.
Sin/salvation. Sin: disobedience, harming others. Jesus died to protect us from God's anger. Salvation is after death, when God can be loving.
Heaven/Hell. Heaven: in bliss with God and Jesus forever. No outsiders.
Hell: Since c.1500 CE, eternal torment for sinners & non-believers.
Kingdom of Heaven. More glorious afterlife than Jesus' sense of what God wants on earth.
Mystical. Distrust of mysticism. Conformity isolates those who have inner experiences.

Modern

Since the Enlightenment, 500 yrs. (Fowler Stage 4, Gebser Mental)
Key words: Rational, social justice. 15-30% of world population

   Bible. Move away from literal interpretation.
God. God as force as much as a being; in everything, panentheistic.
Jesus The historical Jesus. "The modern church is never quite sure what to do with him.[xv]
Prayer. Move toward meditation, contemplation more than asking favors.
Sin/salvation. Sin: causing harm, injustice, oppression. Rejects the extreme guilt of personal sin.
Heaven/Hell. Cannot be sure of heaven or if life continues after death. We create hell by lack of love and injustice..
Kingdom of Heaven. The Kingdom will be in this world,

not the next. Jesus's core challenge is to love, especially the vulnerable and oppressed.
Mystical. Mystics are hallucinating or delusional. What is real is only what can be observed in the physical world.

Post-modern

Emerging since late 19th century – (Fowler 4,5, Gebser Mental, verging Integral)
Key words: Liberation, relative, pluralism, non-hierarchical.
5-10% of world population.
Bible. Historical, metaphorical, transformational, true rather than factual.
God. Like Modern, panentheistic: God is in everything and everything is in God.
Jesus. Ambivalent views; Jesus may be seen as too exclusive or all-inclusive.
Prayer. Many acceptable forms. Everyone has ability to send healing energy.
Sin/salvation. Sin: Not living up to God-given potential; systems of oppression. Universalism.
Heaven/Hell. Death as mystery, afterlife probable, possible reincarnation not resurrection. Authentic Self is eternal and never dies but continues to grow.
Kingdom of Heaven. As in Modern, the goal of life is to bring the Kingdom in this world, not to be saved in the next.
Mystical. Big change. Postmodern thrives on spiritual exploration, embraces transpersonal ideas. Jesus as Wisdom teacher, Spirit bringer.

The same pattern of consciousness that operates in individuals is evident as the systems have progressed over time, moving from a tight circle to greater inclusiveness, from literal meanings to metaphor, an increasing openness from concrete to abstract thought. This is a natural progression, not a deterioration.

---

[1] Johnson, Jeremy. *Seeing Through the World: Jean Gebser and Integral Consciousness.* Seattle: Revelore Press, 2019, chapter 1.
  [i] Sutherland, Cherie. *Transformed by the Light.* New York: Bantam, 1992.
[ii] https://redfrogcoaching.com/uploads/3/4/2/1/34211350/ken_wilber_introduction_to_integral.pdf
[iii] American Psychological Association. https://www.apa.org/topics/trauma/index.html
[iv] Peterson, Jordan B. *Maps Of Meaning: The Architecture Of Belief*, Abingdon, UK: Taylor and Francis Books, 1999. 228
[v] Peck, M. Scott. *The Different Drum.* New York: Touchstone, 1998. 187-203.
[vi] https://www.Mc.Vanderbilt.Edu/Root/Pdfs/Reynolds/Brief_Summary_of_Stages_of Selfhood_and_Faith_ Development.pdf
[vii] https://gebser.org/jean-gebser-bio/
[viii] *Ibid.*
[ix] https://gebser.org/jean-gebser-bio/
[x] Gebser, Jean. *The Ever-Present Origin.* Athens, Ohio: Ohio University Press, 1986. 159
[xi] Gebser, Chapter 3: The Four Mutations of Consciousness.
[xii] Wilber, Ken. *Up from Eden: A Transpersonal View of Human Evolution.* Boulder, CO: Shambhala, 1983. 22
[xiii] *Ibid.*
[xiv] Gebser, 46.
[xv] Smith – Integral

CHAPTER 6.

ARCHETYPES: IMAGES AND FORMS

IMAGES IN SPIRITUAL EXPERIENCE: 'WHAT'S WHAT'

There was the year my husband and I, both teachers, lost our live-in Scottish au pair two weeks before school started, and in a panicked flurry hired a non-English-speaking Columbian woman to do childcare for our two little girls. My husband was fluent in Spanish and could speak with her; I could not. That was the year I discovered how universal it is to draw pictures in order to navigate a foreign language. I also discovered the power of other images.

Maria, who was probably in her forties, had been a factory worker in a town outside Medellin, but she loved children and, one supposes, had thought this might be an adventure. Her agency, however, had obviously done nothing to prepare her or, for that matter, us. Maria, had spent her entire life in a tropical rain forest climate and was utterly uninformed about what to expect in upstate New York. The first clue was the refrigerator. Maria would not open the refrigerator door until she had wrapped a scarf around her face, covering her nose and mouth; cold air, we discovered, could kill a person. In October, she became almost frantic, thinking she had arrived at the end of the world, because all the trees were dying, their leaves falling off, and what would happen when the trees were all dead? In November there was snow, and the plant life had also died. Then she talked with a Columbian acquaintance in the Bronx, who had a story about an ice storm. We did not realize at first the terror Maria was living with, her image being the house entirely encased in an impenetrable layer of ice, freezing the doors and windows so no one could go in or out. And what if this happened while we were at school, leaving her frozen inside with two little girls and no way of knowing how to run the house or get food? The images of her imaginary weather were beginning to kill her. By January we had arranged with the agency to find her a position in a southern state. My sense of guilt has never quite gone away. Images create their own reality!

Ultimately, too, within any rite or dream or non-ordinary experience the combination of images and prior knowledge will generate its power. In NDEs and similar experiences, images are the deeply rooted sensory *things* to which the person makes an emotional response. As I had no context for the Yin-Yang circles, Maria had no context for northern hemisphere weather. For both of us, as for anyone trying to shape a life with a badly understood image, adequate information makes all the difference.

After an altered state experience, with the first waking consciousness, comes the telling of the event, an organizing of words and images, creating a narrative if only for oneself. This story of what happened is an interpretation of the images. The experience itself, which was temporal, is over and gone and irretrievable, remaining in memory as meaningful images but gone in presence like any deceased loved one. The narrative—the verbal account and description of images—is not the experience but its depiction.

If we are still operating out of the Old Filter (as I was), we automatically assume a narrative based on traditional understandings—"I think I was dead, and it was pleasant, so I was in heaven"; "I was falling, so it must have been going to hell." "Predestination." That is all we can do, drawing from our cultural resources. When we are tied to the physical world view of *place*, we are tied to whatever we think belongs there: spiritual entities are named according the expectation, even though they may not identify themselves. Kind, loving, transcendent stranger appears in the NDEs of a Christian and a Hindu: Christian: "Jesus came and walked with me"; Hindu: "Krishna came to help me." Without such familiar connections, I was lost.

Where do the images come from? Do our brains make them up in our sleep? Are they sent from somewhere outside our brains—and by whom? In the majority of experiences, what people report will be connected to what they have internalized from their culture's beliefs—angels, demons, loud noises interpreted as the cacophony of hell or choirs of the blessed. The source of the images will be identified according to the same background,: a perspective as described by religion, secular system, or state of confusion.

From the natural psychological perspective, images in dreams and spiritual experiences come from archetypes, which are understood as being shared across time and cultures, with near-universal meanings; some have more than one.

Too often, the people most needing to understand their images have limited information about them, so they take them at their first impression, literally. Unfortunately, we cannot go back to the experience and look a second time.

An interesting note about the possibility of cultural bias in interpretations is that tunnels have been so frequently mentioned in Western NDEs, investigators expected them to be common universally. As accounts were gathered from Pacific and hunter-gatherer communities, and from remote locales, the early reports were that tunnels were absent. However, what was sometimes included was description of a darkness and emerging from something like the calyx of a flower, or a hollow stem, or the whorl of a shell. Western reporters continue to say there are no tunnels in those experiences.[i] How, I wonder, could they not notice that a hollow stem and natural forms like it are shaped much like a tunnel?

We are left struggling through our inescapable cultural biases to interpret accurately. Do we—or the many experiencers—recognize fire only as torment, or can they access the idea as the presence of the sacred, as purification, or, as with the phoenix, a rebirth? Can they see beyond the terror of a horned creature as Satan to recognize its ancient history as a guide, a mentor? Or know a snake as an honored sign of renewed life, as wisdom? Do they even know about death as a sign of an ending preparatory to new life rather than as their own extinction? And the Void—how many Western people are aware that in other traditions the Void is not the hopeless bottom but the ultimate highest spiritual destination?

In all these cases and countless others, people's minds have been filled with half-truths and even lies about the language of spirit, which is its images. No list of nouns can be applied like WD-40 to fix every stuck interpretation, but some basic literary vocabulary helps.

A fascinating alternative to assigning a quick interpretation to archetypal images comes from James Hillman's archetypal psychology. This may take greater patience than most of us have, and perhaps it requires a therapeutic mentor, but his question is, can we simply sit with the images, without labeling them or leaping to conclusions about their meaning?

Can we simply stay with them until they are familiar and broken in, while we learn to hear what they have to tell us? (I suspect this may be what I have done, though a fifty-year integration span is not recommended.) The images are a 'who's who' of our subconscious.

ARCHETYPES: THE 'BUILT-INS'

Archetypes are sometimes images and sometimes ideas, but they always point to ideas. Psychologist Carl Jung did not invent archetypes, which have been recognized as patterns for millennia, although he coined the term and relied on them heavily in describing the workings of the human psyche. Essentially, archetypes represent forms—patterns or principles or types—of individuals, relationships, concepts, and objects which have appeared across human history and have emotional resonance with humanity. They could be called "Loose Templates of What Things Can Be Like."

The word "archetype" has its origins in ancient Greek, with Plato. The root words are *archein*, which means "original or old"; and *typos*, which means "pattern, model or type." The idea is that an archetype represents the founding form of the image or idea, and that what we experience of it in the world is an example of that idea.

Listen to people trying to explain their NDE or STE to someone who has no idea what they're talking about. What we hear will be layers of comparisons. We hear this again and again in descriptions from antiquity: "It was like…and it was like…and it was as if…and it was in the way of…" The experience was ineffable, beyond description, and so our attempt to capture it is by piling it up, layering layer after layer of images, hoping to catch just the right one. This happens with the attempts to define archetypes, too.

Archetypes are neither individual nor are they images; they are collective, universal forms across cultures and generations, and transcendent, beyond conscious perception.[ii]

"Jungian archetypes are defined as universal, archaic symbols and images. They are the psychic counterpart of instinct ... a kind of innate, unspecific knowledge, derived from the collective unconscious, the sum total of human history."

Wikipedia, quotes Jung: the most far-fetched mythological motifs and symbols can appear [spontaneously] at any time... These "primordial images" or "archetypes," as I have called them, belong to the basic stock of the unconscious psyche and cannot be explained as personal acquisitions. Together they make up that psychic stratum which has been called the collective unconscious.[iii]

Psychoanalyst Lionel Corbett: "[The traditional idea is] that we are influenced by little understood powers, which the ancients called gods or spirits, and we call archetypes."[iv]

Stanislav Grof: Archetypes are timeless primordial **principles** underlying, informing, and forming the fabric of the material world... The concepts of Heaven, Paradise, and Hell can be found in many cultures of the world, but the specific form. . . varies from one instance to another.[v]

Charles Tart: Archetypes are particular biologically inherited psychological structures that can, under the right circumstances, emerge and dominate consciousness because of the high psychic energy residing in them. [Altered states of consciousness] frequently facilitate the emergence of archetypes."[vi]

In other words, of which we are using rather a lot, they constitute the building materials of our continuing experiences.

All that layering! The very fact that archetypes are so difficult to pinpoint tells us they will be complex in their operations also. The thing is, archetypes have no image themselves; the forms are empty until filled with what a human mind puts into them for its own expression. Further, they are not specifically detailed but more like an idea the result leans toward.

They are expressed by the mind but do not come from the mind; they exist, albeit invisibly, in the universal aspect of the deepest psyche and populate the collective unconscious; in other words, we are all connected to it. For example, Shadow, Animus/Anima, Wise Old Wo/Man and finally the Self, are names that Jung gave to archetypal personifications of (unpersonal) unconscious contents which seem to span all cultures.[12]

ARCHETYPAL PURPOSE: THE SAME KIND OF LANGUAGE

One of my favorite teaching sayings is, "Stories are the baskets in which we carry our meanings." We could probably say the same about archetypes. ("Probably" because theorists have field days nit-picking that kind of statement about archetypes.) It's true, though: meanings bond like pond algae on archetypes, which is why they are so closely associated with myths, which of course are stories that stride like giant steps across millennia.

It is not only around campfires that listeners get to hear the myths of their people, though it helps to hear them in the dark. In fact, the *Star Wars* website points out that "Myth is often something experienced unconsciously by a collective."[vii] All those audiences, huddled in the darkness to share a story!

This is a particularly rich time for myth, at least in Western film making. It should surprise no one that the work of the great mythologist Joseph Campbell was hugely influential for George Lukas in writing the films in the Star Wars series. There are the Tolkien *Rings*, and Harry Potter as series, and individual releases of Madelaine L'Engle (*A Wind in the Door*), C.S. Lewis (*Narnia*) and in all of them, great plots and dazzling special effects help carry the archetypal myths to audiences who simply thought they were seeing a great movie.

Nevertheless, the power resides in the experience equally as much as in the understanding."

This is the built-in strength of archetypes doing their work. The *Star Wars* website says, "Even the creators themselves can be part of this collective unconscious. Composer John Williams was in the audience for one of Campbell's lectures at Skywalker Ranch and commented, 'Until Campbell told us what *Star Wars* meant ... we regarded it as a Saturday morning space movie.'"[viii]

Over time and changing cultures, the plots shift, the names and faces differ, but if the scripts are well written (and sometimes even with disasters), archetypes keep speaking their age-old content about human behavior and emotional conditions, looking as far back beyond ourselves as the eroding lines on cave drawings. As the web site observes, "And as scientific discovery continuously redefines our understanding of the cosmos, certainly our mythical perspective must change as well." What the archetypal stories give us is pattern and dimension.

The individual caught by a major archetype will not forget the instance. Its power derives from its universality and its endurance throughout generations of human lives and psyches. Joseph Campbell describes the result as seen in the myth of Ariadne, daughter of King Minos, turns to Daedalus, architect of the dreadful maze which houses the Minotaur, to aid the hero Theseus, who must go into the maze and return.

Daedalus simply presented her with a skein of linen thread, which the visiting hero might fix to the entrance and unwind as he went into the maze. It is, indeed, very little that we need! But lacking that, the adventure into the labyrinth is without hope.

...The flax for the linen of his thread he has gathered from the fields of the human imagination. Centuries of husbandry, decades of diligent culling, the work of numerous hearts and hands, have gone into the hackling, sorting, and spinning of this tightly twisted yarn. Furthermore, we have not even to risk the adventure alone; for the heroes of all time have gone before us; the labyrinth is thoroughly known; we have only to follow the thread of the hero-path. And where we had thought to find an abomination we shall find a god; where we had thought to slay another, we shall slay ourselves; where we had thought to travel outward, we shall come to the center of our own existence; where we had thought to be alone, we shall be with all the world.[ix]

And we find ourselves standing together, all of us and Luke Skywalker and Theseus, Leia and Ariadne, telling stories everyone understands. They are the stories of myths and of exceptional human experiences intimes after times, with plots differing but speaking the same kinds of language.

HOLOTROPICS AND ARCHETYPES: REAL FORMS IN REAL TIME

In more than fifty years of medical practice and psychedelic research, the Czech (now U.S. citizen) psychiatrist Stanislav Grof has amassed a mountain range of evidence pointing to the human psyche as a repository of every imaginable human emotional experience, from the blissful to the violent, the loving to the murderous, the boring to the grotesque. This is what Carl Jung meant by the collective unconscious, an enormous in-house source of archetypes in the depths of the psyche.

Grof came to the U.S. in 1967, already well known for his research on psychedelic drugs and their effect on the psyche. When the government banned psychedelics, Grof and his wife, Christina, developed breathing techniques and support activities which could produce non-ordinary consciousness experiences very similar to those with LSD. Carefully supervised, the experiences had therapeutic value. The Grofs termed their unique approach "holotropic therapy."

What they discovered was that in the course of hours-long experiences of hyperventilation and special music, clients would find themselves accessing deep and usually locked-off areas of the psyche, in contact with powerful archetypal forces of astounding meaning. Over multiple sessions a client would progress through experiencing a series of biologically intense, naturally arising archetypes of the birth process. Richard Tarnas describes it as "a highly charged life-and-death struggle with the contracting uterus and birth canal, and culminating in an experience of complete annihilation," followed by sudden "global liberation" perceived as both physical birth and spiritual rebirth. Tarnas, who was director of programs at Esalen for ten years, says that "In terms of therapeutic effectiveness, Grof's was by far the most powerful. There was no comparison."[x]

Any large breathwork session might include individuals displaying connection to a myriad of archetypal themes, from serene to blissful heavenly visits, to tormented encounters involving monsters and insectoid presences, violent dismemberment, scenes of worldwide warfare and destruction, entrapment and despair. In one pattern, clients instinctively reproduce the stages of physical birth and the emotions associated with each stage, which often reveal profound reverberations occurring in the client's psychological and emotional development years later. In other words, the archetypes activated in holotropic and other altered state experiences can explain the source of even the most grisly scenario, including hell.

Reading Grof led me to the observation that as the distressing holotropic sessions were seen to originate as developmental steps to be worked through en route to personal growth, so distressing NDEs could be understood the same way. A hellish NDE was not limited to an after-death state, nor was it, in this context, dependent on one's behavior or character; it could as well be understood as an experiential pothole rather than a vindictive punishment by an external power source. That perspective opened the door to see its meaning as something better understood and integrated than eternal physical torment. It was a major 'what if?' about the old cultural doctrine. What if a distressing NDE occurs not to enhance guilt or unworthiness but as a means of overcoming them! And what if it might sometimes even be not divine vengeance but happenstance!

No wonder these altered states leave us quaking. Trying to make sense of them in literal terms is futile. What I hope this book will convey is that it is possible to understand the concept of hell and the difficult aspects of any NDE in ways that challenge rather than destroy. They are messages, not punishments.

This approach does not lessen the immediate terror or frustration of the experience but removes its vindictiveness. One has still to deal with the monsters, who can now be understood as something engaging us. What are they saying? What do they want? What are they asking?

Suddenly I could see that we might even learn to prepare for afterlife in a Western tradition, parallel in intent to the Tibetan Book of the Dead, like the medieval *Ars Moriendi*, the Art of Dying. I could see that we can dissociate a hellish NDE from the notion of eternal punishment and use it as an engine of spiritual growth in our own terms, from within a religious tradition or not. In secular terms, the same process would be true. Grof has written:

Instead of useless pieces of knowledge, the data about the hells and heavens can prove to be invaluable cartographies of strange experiential worlds which each of us will have to enter at some point in the future … Avoidance and reluctance to surrender are considered two major dangers. These mythologies and concepts of…heaven and hell…are an intrinsic part of the human personality that cannot be repressed and denied without serious damage.

Notice that the two major dangers Grof mentions are exactly what our culture displays relentlessly: avoidance and reluctance to surrender!

THE 'IMAGINAL' WORLD: REAL BEYOND IMAGINARY

How many times, when talking about an altered state of consciousness, have we heard, "Oh, that's just your imagination! You—or your unconscious—made that up." But we who have had such experiences *know* empirically that there is a difference; we did not participate in the creation of this event, not the way we would if someone said, "Imagine you're at the beach." That would be ordinary imagining, an *imaginary* beach. No, something about these experiences simply arrived all on its own. It is that *something* which makes them feel "realer than real."

The cosmology of Western Enlightenment, of capital-S Science, has essentially two dimensions: real and imaginary; there is nothing else. Our NDEs and STEs, therefore, must be one of the two. Often that is not a clear choice. Of course the experience is real…as an experience. But is it real in the physical world? And if it is not, does that make it an unreal experience, especially if it feels like the realest event of our whole life? Is it "fake news"?

Small children often have an imaginary friend, who is experienced by the child as real—parents know!—real enough not only to have a unique name but as a presence needing a place to sit, a seat at the table, a pillow on the bed; one must wait for the friend to come along before closing the door. The friendship may continue for several years, even until the child is well into school age. (One such friend in our family was named Kacky-Rootl, and everyone still living remembers her, though that was seventy-some years back.) This is the age when children may carry on conversations with other invisible presences, often identified by the child as a deceased grandparent or other relative who comes to play with them. They may report seeing fairies in the back yard or other mythical beings. During this same age span, some children will surprise a parent by mentioning "my other mother," or explain how they know something "from before I was born this time." Such conversations cease and may be denied by the time the child is seven or so, because they have learned that in our ways, there is real and there is imaginary; there is no real in-between.

We have not always had such cold and clear-cut lines. For most of human history, there was room for others—the mythical beings and recognized things-seen-out-of-the-corner-of-an-eye, the dreams hinting of premonition. With the coming of hard-edged science, that changed. The most highly valued attitude in the sciences is objectivity, the absence of emotional bias, reflecting the Newtonian image of the universe as a great machine operating with cool impersonality according to observable (countable) rules. All very clean and unfussy. It made sense, then, that in the early 20th century, when practitioners wanted the newish field of psychology to be accepted as scientific, they would move toward objective principles and practices emphasizing systematic and methodological approaches. The result: behaviorism, cognitive psychology, biological psychology. Like the hard sciences, these approaches are compatible with measurement and visible evidence; they stress observable behaviors, not the hidden unconscious.

We have lost much of what has been called our "enchantment," our ability to resonate with the regions formerly held by poetry, mystery, and the imaginative realms. Mysteries today are mostly solved by policework, not spirituality.

There are still some areas which have not been stripped bare. Anyone dealing with archetypes or spirituality, or with the paranormal or the psyche, knows that there are in fact 'things that go *bump* in the night.' Jungians refer to the world of the archetypal figures and realms as 'imaginal,' making a distinction from imaginary products of the individual human mind. Some version of this view is shared not only in Jungian circles but in many of the world's non-Western religious systems, where the dichotomy of 'real' and 'imaginary' is also absent or at least muted

IMAGINAL: *MUNDUS IMAGINALIS*

Not everyone agreed with the behaviorists. Other psychologists considered those approaches cold, reductive, materialistic, and merely literal. Among those sharing a belief in the importance to psychology of psyche as well as behavior was one of the seminal thinkers of the 20th century, the French Islamist Henry Corbin.

A professor of Islamic Studies at the Sorbonne and a Christian theologian, Corbin was one of the century's major scholars of Sufism and Persian mysticism. His contribution to psychology was his introduction of the concept of *Mundus*

Imaginalis, the imaginal realms, into contemporary thought. That work provided a foundation for James Hillman's archetypal psychology and influenced thinkers and artists worldwide.

The Sufi tradition avoids the problem of rigid dualism and its real/unreal limitations by assuming the existence of a territory between the physical, sensory world and the spirit world. This intermediate world has its own consistent topography but is also constantly influenced and shaped by the physical and the spiritual worlds. Corbin coined the term *Mundus Imaginalis* to explain this third world to Westerners. The *Mundus Imaginalis* is something like the Christian heaven; it's the part of reality where archetypes exist; it is peopled by beings, including angels.[xi]

The imaginal realms, in Corbin's articulation (lightly tinged with Islamic mysticism), have "three universes, or, rather, three categories of universe." There is our physical, sensory world, the world of phenomena, which includes both our earthly world governed by human souls and the sidereal star universe governed by the 'Souls of the Spheres.' There is a suprasensory world of the Soul or Angel-Souls, in which there are the mystical cities of Sufism and, one assumes, other mystical traditions. Third, there is the universe of pure archangelic Intelligences. Corresponding to these three universes are three organs of knowledge: the senses, the imagination, and the intellect, a triad corresponding to the triad of anthropology: body, soul, spirit. Notice Corbin's view that imagination and intellect function as *organs*.[***xii***]

We can see immediately, says Corbin, that we are no longer reduced to a cosmology limited to the empirical world, on one side, and the world of abstract understanding on the other.

Between the two is placed an intermediate world... the world of the Image, *mundus imaginalis*: a world as ontologically real as the world of the senses and the world of the intellect, a world that requires a faculty of perception belonging to it, a faculty that is a cognitive function, a noetic value, as fully real as the faculties of sensory perception or intellectual intuition. This faculty is the imaginative power, the one we must avoid confusing with the imagination that modern man identifies with "fantasy" and that, according to him, produces only the "imaginary[xiii]

Contemporary mystic Cynthia Bourgeault has observed that although *imaginal realm,* as a term, originates in Islamic mysticism, the idea—really, she says, an archetype more than an idea—appears across all the great sacred traditions. The imaginal realm is a boundary between two worlds, each with its own conventions and structures, unfolding in its own way. Although separated from the visible world, the invisible realms can still be perceived through the eye of the heart, not by mental reflection or fantasy.

'Where the two seas meet' is a beautiful Sufi metaphor to convey the essence of what actually happens here. The imaginal realm is a meeting ground, a place of active exchange between two bandwidths of reality…[xiv]

It's in that realm that NDEs and STEs and all have their existence. The difference between 'imaginary' and 'imaginal' is, as my father would have said, "Broad as the side of a barn!" In other words, it's not a trivial difference, especially when considering altered states of consciousness like ours. The distinction is somewhat akin to the difference between voluntary and involuntary biological systems, with 'imaginative' being related to ordinary waking consciousness and 'imaginal' requiring special access, such as via an altered state.

Imaginal realm. There is no term like this in Newtonian physics, though there is in quantum. Sometimes we are required to shift domains.

ACTIVE IMAGINATION: WHERE A REFLECTION LIVES

What makes the imaginal world function deliberately is what Corbin and Jung called active imagination. You know the saying, "Some achieve greatness; some have greatness thrust upon them"; substitute "imaginal realms" for "greatness." Those who have had a transcendent NDE or SCE have felt the imaginal world thrust upon them. For others, the use of active imagination can, with practice, achieve something of the same insight.

We are not talking about surface level imagination, the kind that lets you instantly imagine your favorite ice cream or imagine being in the house where you grew up. This type invites the deeper psyche to participate, which makes it both powerful and sometimes dangerous.

Note: Inviting the deeper psyche is contraindicated when there is any hint of risk to mental health. Persons dealing with schizophrenia and other dissociative conditions should not undertake any active imagination exercise except under professional supervision.

Active imagination is a method for visualizing unconscious issues by letting them act themselves out. At its simplest, active imagination is doing with consciousness something rather like what we do when we unfocus our eyes: we let awake awareness slip away without letting go entirely. What is then possible is that as the awakeness softens and relaxes its defensive grip on physical surroundings, the psyche can relax and begin to open, giving access to contents which are ordinarily hidden.

Some people focus on a specific image, perhaps something seen in a dream, keeping the image the only object allowed into their concentration, waiting to see what will happen with it. Some use automatic writing. Others engage a character, again maybe from a dream, or a biblical or literary character and ask a question to which they really want an answer. (This can be done entirely mentally, or while sitting at the computer, letting the awake awareness slide away, then typing a question and taking notes—or typing quotes—from the response.) The point is to move in toward the psyche and engage with it, as Stan Graf's clients do with holotropic breathwork. Surprising results are common.

In imaginal terms, the "where" of the responses is like the "where" of a mirror's reflection. There is a genuine reflection, real information, but no physical place where they reside. Corbin called the mirror-like function "an epiphanic place"—a place of epiphany—a sudden, intuitive insight into the reality of something, usually initiated by something simple or commonplace. The active imagination, he says, is

... the preeminent *mirror*, the epiphanic place of the Images of the archetypal world; that is why the theory of the mundus imaginalis is bound up with a theory of imaginative knowledge and imaginative function... It is a function that permits all the universes to symbolize with one another or exist in symbolic relationship with one another...

It operates much like the *tertium non datur[xv]* concept of seemingly endless tension, an irreconcilable conflict in which frustration at finding no answer builds and builds, neither of the dualistic responses being acceptable, until suddenly a third option erupts, a previously unseen and new resolution. What Corbin observes is that the cognitive function of the Imagination escapes from the dualistic dilemma of today's rationalism, which gives only an "matter" or "spirit" answer (real or imaginary) and permits the establishment of a creative way, a "rigorous analogical knowledge" *without resorting to banal dualism.*

Corbin's work is significant for our purposes not only because of the term *imaginal* but because of his interweaving of psychology with issues in visionary spirituality. His profound scholarship on Islamic mysticism in a framework of understanding the unity of the religions of the Book: Judaism, Christianity and Islam led to a passion to free the religious imagination from fundamentalisms of every kind.

*As Corbin and Hillman have emphasized, literalism is one of the chief strategies of the fundamentalist and totalitarian mind. Literalism encourages the uncritical use of power and coercion because it presumes that reality can be known with certainty. Humility, caution and critical self-awareness are essential for a life in tune with imaginal realities.[**xvi**]*

The perceptiveness of Corbin and Hillman helps explain why fundamentalists of any type, secular as well as religious, have such difficulties with altered state experiences. The reason is that tradition (doctrine, law) prizes order and certainty, whereas these experiences thrive on the flexibility of epiphanies and chaos. NDEs and STEs and other mystical events do not operate by set doctrines and standards. That is why all major religions have originated in the profound altered states of consciousness of charismatic individuals (Mohammed, Jesus, Lao Tse, the Buddha) which led people to follow them, but it is the organizational beliefs of the followers which formed the religion. See how this puts us squarely at the intersection where the collision of belief systems will be most forceful!

Here is the problem of anyone whose altered state experience did not exactly correspond to what they expected. It is the problem of the Christian NDErs who come back after a beautiful and transforming experience wondering plaintively, "But...was that biblical?" Or the convinced atheist who comes back shaken because, as Ken Ring says, "He doesn't know what he saw, he only knows it wasn't supposed to be there." It is the problem of the Jewish mother wondering about a totally loving entity she referred to dubiously as "that

Jesus look-alike." This is the problem of the countless families deeply troubled theologically when they hear that a suicide attempt was not punished, or that a woman who had an abortion was warmly greeted by Jesus. It is why there will be claims that an NDE was satanic because the drag queen cousin was welcomed. For an authoritarian (fundamentalist) observer guarding a narrow range of beliefs, only a tight rein can be trusted; anything else (difference) introduces change, which is heresy. The problem is that one reality does not match the other.

What the great majority of American adults know about interpreting religious images could go in their little fingernail, working with ideas remembered dimly from elementary school religious education, which customarily ends around age twelve. Consider what your mind would be like if today you knew about politics or finance or science only what you knew at the age of twelve! For those who approach understanding an NDE or STE with a quick religious explanation, almost as many slam the door on any mention of religious belief. We have, most of us, not been taught how to approach the images and symbols of inner experience.

As I have written previously,[xvii] "What has been largely missing from discussion about these experiences is the explicit acknowledgment that *because* these experiences are ineffable, they have no precise denotation. Like the Sacred, they have many images but no physicality. It is exactly for this reason that the language of science cannot deal with them." Their meaning has been left to others, primarily religion and philosophy.

At a time when our organized faith literacy has dimmed, theologian Walter Wink suggested a fresh perspective: Wink (1986,)

"The new age dawning may not 'believe in' angels and demons the way an earlier period believed in them. But these

Powers may be granted a happier fate: to be understood as symbolic of the 'withinness' of institutions, structures, and systems. People may never again regard them as quasi-material beings flapping around in the sky, but perhaps they will come to see them as the actual spirituality of actual entities in the real world. They are not 'mere' symbols—that too is the language of the old worldview that is passing, for we now know that nothing is more powerful than a living symbol. As symbols they point to something real, something the worldview of materialism never learned to name and therefore never could confront."

We are on our way to confrontation.

LIGHT BRINGERS AND NON-MATERIAL FORMS

And then I came upon this article, "Carl Gustav Jung, quantum physics and the spiritual mind: a mystical vision of the twenty-first century," published in PubMed. Its authors are neuropsychologist Diego Valadas Ponte, General Director of the Portuguese National Stroke Association, and the late Lothar Schäfer, a quantum chemist, Distinguished Professor emeritus of physical chemistry at the University of Arkansas. What advanced mathematicians make of their work is beyond my powers of discernment, but I cannot recommend it highly enough for any layperson who is struggling to understand quantum basics.

By the way in which it describes the world, quantum physics has taken science into the center of ancient spiritual teachings. For example, molecular wave functions have no units of matter or energy. They are pure, non-material forms. The same is true for Jung's archetypes: like the wave functions of quantum systems, they are pure, non-material forms...The discovery of a realm of non-material forms, which exist in the physical reality as the basis of the visible world, makes it possible to accept the view that the archetypes are truly existing, real forms, which can appear in our mind out of a cosmic realm, in which they are stored. Thus, we can confirm here, on the basis of the quantum phenomena, Jung's view that "it is not only possible but fairly probable, even, that psyche and matter are two different aspects of one and the same thing"

> The aspects of the psyche that [Jung] discovered are so profound, that they go beyond the limited concerns of the human psyche, making it possible to think, for example, that the universe itself is conscious and our own consciousness is connected with the cosmic consciousness
>
> By studying the human psyche, Jung discovered **archetypes, the mental properties of the universe**, which Classical physics had suppressed: Quantum physics has now brought them back.
>
> **The facts show us that there is a non-empirical realm of reality, that doesn't consist of things, but of forms.** These forms are real, even though they are invisible, because they have the potential to appear in the empirical world and act in it. They can do this in two ways: they can find consciousness as thoughts in our mind; and actualize as material structures in the external world. Thus, the conscious and empirical world is an emanation out of a realm of mind-like forms, and **quantum physics is a form of psychology, the psychology of the cosmic mind**. In the same way Jung's psychology is also a branch of physics; that is, the physics of the mental order of the universe.[xviii]

This is scientific writing at its purest and clearest!

---

[i] Kellehear, Allan. "Census of Non-Western Near-Death Experiences to 2005: Observations and Critical Reflections," *Handbook of Near-Death Experiences*. Santa Barbara, CA: Praeger Publishers, 2009, 150-152.
[ii] Http://Siivola.Org/Markk U/Uni/Engl_Articles/Difference_Between_Archetypes and_Archetypal_Symbols.Html
[iii] Https://En.Wikipedia.Org/Wiki/Collective_Unconscious

[iv] Corbett, *Religious Function,* 120.

[v] Grof, *Ultimate Journey,* 43

[vi] Tart, Charles T. *States of Consciousness.* New York, E.P.Dutton, 1975, 113.

[vii] Https://Www.Starwars.Com/News/Mythic-Discovery-within-the-Inner-Reaches-of-Outer-Space-Joseph-Campbell-Meets-George-Lucas-Part-I.

[viii] *Ibid.*

[ix] Campbell, Joseph. *The Hero with a Thousand Faces.* Princeton, NJ: Princeton University Press, 1968. 23, 24ff.

[x] Tarnas, Richard. *The Passion of the Western Mind: Understanding the Ideas That Have Shaped Our World View.* New York: Ballantine Books, 1991, 426.

[xi] http://neuroscienceandpsi.blogspot.com/2012/03/Henry-Corbins-Mundus-Imaginalis-Sufism.Html

[xii] *Ibid*

[xiii] Https://www.amiscorbin.com/En/Bibliography/Mundus-Imaginalis-or-the-Imaginary-and-the-Imaginal/

[xiv] Https://Cynthiabourgeault.Org/2018/11/13/Introducing-The-Imaginal/.

[xv] Joshi, Sheila and Barbara Croner. "Tertium Non Datur," Chapter 16 in Bush, N.E., *The Buddha in Hell: Perspectives on Distressing Near-Death Experiences,* 2015.

[xvi] Http://Henrycorbinproject.Blogspot.Com/2009/06/Literalizing-Imaginal.Html.

[xvii] Bush, Nancy Evans. *Dancing Past the Dark: Distressing Near-Death Experiences. 2012, p.* 200.

[xviii] Valadas Ponte, Diogo and Schäfer, Lothar. "Carl Gustav Jung, quantum physics and the spiritual mind: a mystical vision of the twenty-first century." *Behavioral Sciences (Basel, Switzerland)* vol. 3,4 601-18. 13 Nov. 2013, doi:10.3390/bs3040601.

CHAPTER 7.

A RENOVATED STRUCTURE

Integration, for me, has been like geology: mostly a slow-moving trek. Everything takes eons. Every once in a great while, as with Gebser's mutations, there is a leap. In 1997, my fifteenth year with IANDS, a cosmological bombshell (my view) arrived in the form of an article in the *Journal of Near-Death Studies* (JNDS).

For thirty-five years I had been looking at my NDE through the old religio-cultural filter, thinking of my experience as a black mark on my moral character, something entirely negative and perhaps damning in my psyche. At that point, only two people had ever suggested there might be a different way of looking at it; one was Grof, whom I have never met, and the other was the Rev. Peter Grandy, the associate pastor at my church, who told me some traditions consider the Void the height of spiritual attainment, and I might want to read about Buddhism. Otherwise, I have no remembrance that anyone I had encountered at that point, either within IANDS or beyond, had offered any information to counter my bleak view. A few had said they would pray for me.

And then came this article in the *Journal of Near-Death Studies*, "The Mystery of Frightening Transcendent Experiences," by Polish philosopher Mishka Jambor. It was a reaction to several recent articles, one of them mine, which had expressed differing views of distressing experiences. The Jambor article is subtitled, "A Rejoinder to Nancy Evans Bush and Christopher Bache."

The core of her article, based on a study of mysticism, dealt with the very nature of our reality. At the dualistic level, she pointed out, pairs are *required*. If there is Bliss, there must be its corollary, which she calls Abyss. She proposed them as the first elements to emerge from the Ultimate, which lies just beyond as the One. Jambor suggested Bliss and Abyss as "primordial building blocks of the manifest world," underground engines at the emergence of human behavior and experience. The sensory experiences "are glimpses of the Ultimate Reality, colored frighteningly or pleasantly."

Distressing NDEs, she said outright, are as real philosophically as the classic, pleasant NDEs. They are *necessary*.

Jambor asked, "What kind of beings are we, that we can feel in such a profound way that *the feelings alone make for heaven and hell?*

Those are her italics, illustrating the seriousness of her question. "*The feelings alone make for heaven and hell.*" She did not voice the next question, but it leaped at me: *So, if we change the feelings...?* It was the first time I can remember considering the part *feelings might play* about an NDE. This article was the first assurance that I might be on a steady path.

"I congratulate Nancy Evans Bush," she said, "for doing justice to the frightening near-death experience as the depths of spiritual experience as compared with the "heights" of the radiant NDEs." It was a stunning affirmation. A first.

What if? What if she were right? What if we accept the existence of Abyss as essential to Being? What if the minds having "negative" NDEs are not fit only for the slums? Then this is not a trash NDE.

WHY WE THINK THE WAY WE DO

TWO PARADIGMS

Richard Tarnas, in *Cosmos and Psyche,* posits a similar perspective, as in the clamor of major debates in today's Western cultures he sees,

> ...looming behind them two fundamental paradigms, two great myths, diametrically opposite in character, concerning human history and the evolution of human consciousness...those enduring archetypal structures of meaning that so profoundly inform our cultural psyche and shape our beliefs that they constitute the very means through which we construe something as fact. They invisibly constellate our vision. They filter and reveal our data, structure our imagination, permeate our ways of knowing and acting.[i]

The first of the paradigms, he says, "...describes an epic narrative of human progress from a primitive world of dark ignorance, suffering, and limitation to a bright modern world of ever-increasing knowledge, freedom, and well-being." Based on human reason and the emergence of the modern mind, seeing history as onward and upward, its apex the rise of modern science and democratic individualism, this is the paradigm leading to a metaphysical faith in the light, in abundance, in the quest for happiness.

The other paradigm is darker. In this understanding, human history and the evolution of human consciousness are seen as a predominantly problematic, even tragic narrative of humanity's gradual but radical fall and separation from an original state of oneness with nature and an encompassing spiritual dimension of being.

> From profound sacred unity and interconnectedness, the influence of the Western mind brought about a deep schism and desacralization of the world. This development coincided with

an increasingly destructive exploitation of nature, devastation of indigenous cultures, loss of faith in spiritual realities, experiencing itself as shallow, and unfulfilled...In this perspective, both humanity and nature are seen as having suffered grievously under a long exploitative, dualistic vision of the world, with the worst consequences being produced by the oppressive hegemony of modern industrial societies empowered by Western science and technology.[1]

This is the paradigm that recognizes shadow and suffering.

In Tarnas's summary: "They represent two basic antithetical myths of historical self-understanding: the myth of Progress and what in its earlier incarnations was called the myth of the Fall. They underlie and influence virtually all discussions, and constitute the underlying argument of our time...Is history ultimately a narrative of progress or of tragedy?"

Both views, he concludes, are fully valid and yet they are intensely *partial* views of a larger frame of reference which makes a complex, integrated whole. The difficulty for us is

> "to see both images, both truths, simultaneously, to suppress nothing to remain open to the paradox, to maintain the tension of opposites. Wisdom, like compassion, often seems to require of us that we hold multiple realities in our consciousness at once. This may be the task we must begin to engage if we wish to gain a deeper understanding of the evolution of human consciousness—to see that long intellectual and spiritual journey moving through stages of increasing differentiation and complexity, as having brought about both a progressive ascent to autonomy and a tragic fall from unity—and perhaps, as having prepared the way for a synthesis on a new level."[2]

This is where we are now, at the time for that new level which is not yet clear to us. It is dualism that tells us reasoning must be binary—progress/fall, good/bad, light/dark/ bliss/ abyss. But now we see the Enlightenment vision beginning to encounter its own shadow—recognition that both are true leads to the unfolding of a new and more comprehensive understanding. This is the entry—or the access to the entry—of humanity's next level: Gebser's vision, Teilhard's (and Ken Ring's) Omega. In this demanding and complex time, we begin by learning to incorporate all of spiritual experience, the dark NDEs as well as the light, the suffering as well as the joy.

This is where we are now, with foundations crumbling.

ABYSS/NADIR: EMOTIONAL ALCHEMY

In my years of hearing from people emotionally paralyzed or panic-stricken by a horrific NDE, it has become clear that there are few reliable resources. Most Evangelical pastors are stuck in an old theology of Satan and hellfire; mainstream and more liberal clergy have rejected a theology of hell and don't know its history; also, they don't know how to talk about shadow and dark spiritual events; counselors seem to have no clue; and most experiencers don't think about, or can't afford, talking with a Jungian therapist. IANDS groups are often helpful for people whose NDE struggle is back on Earth, but not if their NDE demands taking the dark seriously.

I have been disappointed that, for the most part, commentary in the near-death literature is repetition of the tired assumptions of why it is supposed people get such an experience (they're "mean, unloving, angry, cold, guilt-ridden, depressed, hostile," etc.) and the mantra of "getting what you attract." Are these dismal emotions present because because they are genuinely at the core of the person's psychology or because it's been a hard week at work? For the person who self-identifies as having been "a bad actor' or 'a mean guy,' is there objective evidence to back up that opinion? Since Bruce Greyson and I published our research results in 1992, there has been only one actual study of issues in distressing NDEs

(University of Liege, 2019). That's a long time to wait for new information.

What are we to do with these perceptions and feelings so few want to acknowledge? Whether in the hollow coldness of nightmarish NDEs or the empty, lost feelings and disconnection of coming back—these are real sensations affecting real lives, and they require dealing with. At the very least, how we understand them will affect our movement along the path of discovery. They also offer up treasure. At least in some circles these states of consciousness used to bu called 'desolation' or 'nadir' experiences. I think that is more expressive than simply 'distressing,' as the terms suggest a spiritual dimension. Note to self and readers: We need to find and publicize sympathetic writers.

DESOLATION/NADIR EXPERIENCE

Writing for the Exceptional Human Experience website, its former Executive Director, Rhea White observed:

> Abraham Maslow saw desolation/nadir experiences on the same continuum as peak experiences and as involving an interrelationship between light and dark. In Yin-Yang symbolism, one is constantly becoming more while the other becomes less. It has been my experience (White) that experiences of great desolation may precede peak experiences, though it may take some period of time for this to happen. In a sense, feeling desolate puts you in touch with the depths and heights of self unguessed in ordinary life the same way that a peak experience does.[ii]

Psychotherapist Miriam Greenspan was born to Holocaust-survivor parents in a postwar refugee camp; she is known for her work with trauma. Grief, fear and despair, she says: these are the emotions we most avoid and that we most need to attend to in our time because we reject them most strongly. Suppressed and benumbed, they turn into depression, anxiety, addiction, prejudice, toxic rage, and violence, destructive acts to oneself and others. There are no negative emotions, she says, just unskillful ways of coping with emotions we can't bear.

Although generally devalued in our culture, the dark emotions have a wisdom that is essential to the work of healing and transformation on both individual and collective levels. While conventional wisdom warns us of the harmful effects of "negative" emotions, this revolutionary view offers a more hopeful approach: there is a redemptive power in our worst feelings.

Greenspan argues that it's the avoidance and denial of the dark emotions that results in the escalating psychological disorders of our time: depression, anxiety, addiction, psychic numbing, and irrational violence. Her work shows how to trust the wisdom of the dark emotions to guide, heal, and transform our lives and our world. Drawing on stories from her psychotherapy practice and personal life, and including a set of emotional exercises, she teaches the art of what she terms 'emotional alchemy,' by which grief turns to gratitude, fear opens the door to joy, and despair becomes the ground of a more resilient faith in life. Greenspan argues that it's the avoidance and denial of the dark emotions that results in the escalating psychological disorders of our time: depression, anxiety, addiction, psychic numbing, and irrational violence.

Whereas conventional wisdom warns us of the harmful effects of 'negative' emotions, Greenspan looks at their redemptive power. Trust their wisdom, she says. The way to go behind the lines is to discover positive ways of interacting with unpleasant emotions and spiritual experience. Yes, when the painful emotions can be befriended—though not taken on as boarders—the path leads to transformation and unexpected joy.

A therapist for more than three decades, Greenspan writes:

> There is a shamanism of the dark emotions—a way of maintaining awareness in the midst of the chaos and turbulence of the darker regions of the psyche—that ultimately alters our perception of who and what we are. Painful emotions challenge us to know the sacred in the broken; to develop an enlarged sense of Self beyond the suffering ego... When we master the art of staying fully awake in their presence, they move us through suffering. We discover that the darkness has its own light.[3]

Her book is *Healing Through the Dark Emotions,* published by Shambhala. It includes a highly usable self-help section.

"What would happen if we stopped calling painful emotions 'negative' and had faith that the heart—even in the throes of great pain—can be trusted? 'Personal odysseys through the dark emotions [can be seen] as a path to sacred power.'"[4]

RETURN FROM THE ENCHANTED KINGDOM

Why are we so pasted to the view that the dark emotions are negative, even evil? Somewhere in that glue, I suspect, is the Western conception of an afterlife of torture. Now, I did not particularly believe in hell, and my liberal Protestant denomination decidedly did not teach about a wrathful god who is so angry he would torment even little children forever and ever for flawed theology or anything else. I have never believed in that God or that hell. But then, what was that experience of mine *about?* I had no idea. Nothing I had ever learned about religion, especially my own, gave any insight; the only explanation was hell. Yet the experience had happened and was therefore a real experience, whatever that meant. There was a long way to go.

Today's more contemporary answer does not talk of hell as a place but of hell as an experience. It is an encounter with the deepest, most difficult parts of our own psyche. Here is the situation, quoted from a book I recommend to everyone—psychoanalyst John Ryan Haule's *Perils of the Soul*—

> [Shadow] means our dark unconscious side—everything we left out when we constructed our lifeworld. To enter the domain of the shadow, therefore, is a devastating experience. We collide with everything that seems demonic and destructive. ...Everything we have taken for granted is called into question. The domain of the shadow and the beings that dwell there are inevitably experienced as satanic and abusive, for they pose a monumental challenge to our old world and way of life.[iii]

All those grandparents and parents of our cultural history have told us that disobedience to God—which is usually interpreted to mean every single tiny thing we have ever in our lives done less than perfectly—will take us straight to eternal punishment by a wrathful God. That one frat party you wish you could forget about? Your brother's broken leg from your roughhousing play? The money you 'borrowed' from your mom's purse?

This is the attitude which leads people, in a panic, to scrutinize every corner of their behaviors to see what they think might be considered bad enough to justify that experdience. In other words, they go looking for their sins, thinking that will fix shadow. The underlying issue is that we will be anxiously fiddling around trying to patch our superficial gaps and flaws while the ever-present shadow is waiting for us to see there's another whole realm of consciousness demanding our attention.

Out of my own experience, I say first, be careful. This is likely to be a lonely road. Do not trust shortcuts. Well-meaning people will say things to you that are not helpful, or wrong, or hurtful. There are phonies and crackpots, which is fine as long as you see them for who they are. If you cannot trust a person to keep your truth safe, do not tell them. If your experience flies in the face of your religion's usual teachings, be careful, especially when talking to clergy; be ready to heara strongly unsympathetic response. You can be sure there are other points of view within your faith community (though perhaps not in a sspecific church or narrow denomination) There will almost always be a way to remain true to your faith without compromising what you know from your NDE to be true; read newsletters and blogs, such as the Patheos online library, which covers a vast range of religions and varying belief styles. Visit different worship centers, if you can, and begin hanging out in the online near-death and STE groups until you know who is there and what kinds of things they are saying. There are conservative religious forums and more open ones, and some are secular. Even within these groups, be cautious until you know who's who.

As an example of what can happen, there was this, a few months after I had arrived at IANDS. One afternoon the lead student volunteer and a widely-loved experiencer were hanging out in the office, and the three of us were talking, of course. I had still never mentioned my NDE—the frightening ones were called "negative" in those days—but somehow during that discussion I worked up courage to mention it to these people who were likely to understand. They had both been working hard to understand near-death studies and their developing intuitive (psychic) abilities, and now found themselves unexpectedly confronted by a live subject. They were intrigued by why I would have had an experience so different from the mainstream—my first of a zillion instances of "What kind of people have a negative NDE?"—and began, they said, invoking their psychic abilities to read my aura. (I still knew next to nothing about psychic anything, or auras either, which for all I knew might unscroll like a teletype of my deepest psyche. My 'circles' also had secret knowledge, and now, so did these two.) They squinched their eyes and peered at me and at my head, and speculated about what they were seeing (Entities? Power centers? The Dark Side?) Their interpretation was that there was *something dark* in there, something *heavy and maybe evil.* Their analysis got to the point of, "See, look there...whaddya think of that area...it's so dark!" In their utter fascination, they forgot there was a person on the other side of the conversation, a person who was now badly shaken.

I had been wondering for twenty years about that NDE, and their conversation terrified me. I believe they did not intend harm, but even without intent, they got caught up in their supposed psychic powers, let their egos run, and forgot about restraint, while their unhappy subject did not know enough to stop them. When Ken Ring walked in, he found a sobbing, incoherent executive director and two good friends engaged in serious debate about a real negative NDE.

Know something about your listener. If you are new to "all this," try not to let yourself be carried away or intimidated by the allure of psychic abilities, your own or anyone else's; it's not magic, and is as susceptible to misinterpretation and poor judgment as any other skill. Be smarter than I was: don't let amateurs practice on you! Find other avenues. Take a look at chapter XX for some ideas. And please keep reading.

WHAT DO CONTEMPORARY MYSTICS SAY

Is there evidence that only light experiences can be spiritual? Well, there are these alternatives as quick examples:

- Shamanic initiation including being
- The torture, crucifixion, and forsakenness of the major Western religious figure;
- dNDEs follow the classic spiritual stages: suffering, death, resurrection

And four widely recognized contemporary mystics:

Caroline Myss

What we view as "suffering" may be seen as a necessity. Only since humankind took up residence on the planet have the organic systems of nature been referred to as "disasters." Prior to that, nature's ecological system functioned according to its system of checks and balances, which inspired the Way of the Tao.[iv]

Pema Chodron

Reaching our limit is not some kind of punishment. It's actually a sign of health that, when we meet the place where we are about to die, we feel fear and trembling. A further sign of health is that we don't become undone by fear and trembling, but we take it as a message that it's time to stop struggling and look directly at what's threatening us. messengers telling us that we're about to go into unknown territory.[v]

Connie Zweig

I had believed, with a kind of spiritual hubris, that a deep and committed inner life would protect me from human suffering, that I could somehow deflate the power of the shadow with my metaphysical practices and beliefs. I had assumed, in effect, that it was managed, as I managed my moods and my diet… self-control… Seekers are often led to believe that, with the right teacher or the right practice, they can transcend to higher levels of awareness without dealing with their more petty vices or ugly emotional attachments. It doesn't work.[vi]

Walter Brueggemann

It is my judgment that this [avoidance of the lament psalms (e.g., insistence on positive attitudes] is less defiance guided by faith, and much more a frightened, numb denial and deception that does not want to acknowledge or experience the disorientation of life. The reason for such relentless affirmation of orientation seems to me, not from faith, but from the wishful optimism of our culture. [vii]

How much did their views make us want to argue back?

Wherever we resist most strongly, that's where we need to look at our thinking.

---

[i] Tarnas, *Cosmos and Psyche*, 10.
[ii] Rhea White, Http://Www.Ehe.Org/Display/Ehe-Page.Cfm?Id=4.
[iii] Haule, 195.
[iv] Myss, Caroline. https://www.myss.com/inspiration-and-expectation/ August 2014.
[v] Chodron, Pema. *When Things Fall Apart: Heart Advice for Difficult Times.* Boulder, CO: Shambhala, 14.
[vi] Zweig, Connie. *Meeting the Shadow: A Consciousness Reader.* Los Angeles, CA: Tarcher. 1991, *xiv*.
[vii] Brueggemann, Walter. *Message Of The Psalms.* Minneapolis, MN: Augsburg Fortress, 1985. 51-52.

Notes

1. *ibid.*
2. Tarnas, Cosmos and Psyche, p. 14.
3. Miriam Greenspan, Healing through the Dark Emotions, 27.
4. *ibid.*

CHAPTER 8.

CHAOS AND FRACTAL BEING

There is a point to all of this. I have mentioned that my prevailing questions have been not so much about doctrinal theological issues or, more directly, about a fear of hell, as they are about 'How does this place work'? Meaning...the world, the universe, probably as revealed by their nature. Behind the thin disguise of that seemingly impartial inquiry is a gaggle of philosophical and theological questions about God, identity, and Being, none of which seems approachable. But it turns out, I am able to get very close to answers by going a route through some sciences, including the social sciences and now mathematics.

The outline of that route is in this book's table of contents, although the road is not as linear as the page. I remember saying once to Leslee Morabito, after we had worked together for a couple of years, that learning about "all this" was like finding a thread dangling from the sky—and I would reach up as if to gather one—and then another, and another; and if one could only get enough of them braided together, there would be understanding one could work with.

This book is a partial gathering of my threads, most of them, as it turns out, having to do not specifically with religion but with a panentheistic "all this." embracing a range of topics from the normal to the paranormal, from the new science to the old science (even so-called "pseudoscience"), from normal consciousness to alternative consciousness... each weaving its thread into the greater cord that leads everything back home.

What is true about how the world works? For a great many experiencers, an NDE or STE or EHE exposes some deeply surprising truths they have never seen before, truths which are unquestionably real and enduring. The deepest concern when they come back occurs around the shock issue, when the powerful reality of their altered state of consciousness (ASC) contradicts the expectations of their lifelong previous worldview. Before, they *just believed* the religious or atheistic teachings they had been taught represented True Reality; but now they *know*...what? Did that realer-than-real experience mean Satan was trying to ensnare them into false beliefs and a sure hell? The devoted atheist, the believer in scientism, *just believed* the notion of spiritual experience was made-up nonsense, but now, after an NDE...what? Their lifelong reality has *just collapsed*.

What to do with a fallen model? One possibility is to refuse to accept the fall. Many experiencers with a background of rigid belief cannot bring themselves to tolerate the idea of its being wrong.

A militant atheist, confronted by a clear memory of a transcendent experience he does not want to believe, can take refuge in the idea of psychosis: "That was a momentary psychotic break. A bit of insanity, a lack of oxygen. I'll be okay now."

A religious extremist, stiff with doctrine which did not show up in the experience, projects the event onto demonic forces: "That was the work of the devil!"

For these individuals and those who agree with them, what they want is the security of a stable, unchanging reality, universal truths and unvarying laws. Those arguments go very far back in human philosophical history. But as even science has been discovering, reality is more naturally *inconsistent*.

CHAOS

The calm, dependable reality those folks long for was for many years considered the normal way of things. For most of the twentieth century, conventional wisdom in medicine held that an orderly and machinelike system were normal, and that aging and disease arose out of stress. Everyone agreed, chaos disrupts normal peace and is to be avoided at all costs.

Early in the 1980s (those were busy years!), researchers at Harvard and elsewhere expected that chaos would be most apparent in diseased or aged systems, based on both their own intuition and available scientific data. What they discovered was that healthy heart rates fluctuate considerably even at rest; in fact, the heart and other physiological systems may behave most erratically when they are young and healthy. Counterintuitively, too-regular heart rhythm often accompanies aging and disease, sometimes even sudden death. *Chaos is a normal feature* of other components of the nervous system also.

The chaos which research was finding is not what we mean when something is utterly disorganized and random. The disruptions in heart rate and other system reports were typically small, always unexpected, and could have huge, far-reaching effects. In a nonlinear system, because the change of an output is not proportional to its input, some aspects of the system's behavior can appear to be counterintuitive, unpredictable or even chaotic. The "butterfly effect" is an example, the idea that with precise location and weather conditions over sufficient time, the fluttering of a butterfly's wings in one continent can lead to a hurricane thousands of miles away.

Over a longer period of time, the researchers found, the disruptions showed a pattern of occurrence; they were not entirely random, though again, always unexpected.

Chaos is the science of surprises, of the nonlinear and the unpredictable. It teaches us to expect the unexpected. While most traditional science deals with supposedly predictable phenomena like gravity, electricity, or chemical reactions, Chaos Theory deals with nonlinear things that are effectively impossible to predict or control, like turbulence, weather, the stock market, our brain states, and so on.

These phenomena are often described by fractal mathematics, which captures the infinite complexity of nature. Many natural objects exhibit fractal properties, including landscapes, clouds, trees, organs, rivers, and many of the systems in which we live exhibit complex, chaotic behavior. Recognizing the chaotic, fractal nature of our world can give us new insight, power, and wisdom. For example, by understanding the complex, chaotic dynamics of the atmosphere, a balloon pilot can "steer" a balloon to a desired location. By understanding that our ecosystems, our social systems, and our economic systems are interconnected, we can hope to avoid actions which may end up being detrimental to our long-term well-being.[i]

> "Nonlinear chaos," says a leader in the field, "refers to a constrained kind of randomness which, remarkably, may be associated with fractal geometry. Fractals consist of geometric fragments of varying size and orientation but similar shape. Because of its ever finer detail, its length is not defined. In the human, fractal-like structures abound in networks of airways, blood vessels, nerves and ducts. The most carefully studied fractal in the body is the system of tubes that transport gas to and from the lungs. [ii]

How does all this speak to the struggle to understand what seems to be the chaos of an NDE? One approach is to think of the change as potential growth rather than disorder. This is obviously a variant of seeing a challenge in place of punishment, a potential positive rather than the doom in the common wisdom.

In other words, one can take refuge in chaos theory, realizing that growth is a natural happening, that beliefs are not static, that they develop across time just as childhood does, and disruptions can be healthy. This is a major step forward for many beliefs! Beliefs are not locked into concrete. There is more than one way to think about anything. Some ways seem more solid than others, and more to our personal liking; sometimes we can make our own choices. With a firm hold on this idea of growth it is then possible to turn and begin applying it to your own thoughts. Be ready for surprises. If you are worried about losing your faith, you can relax. You won't lose it, though you may have to allow some new ideas and conflict with people who are not yet ready for more growth.

What may be most interesting about stages in general is not so much their specific content as their *pattern*. Look at the pattern of the three systems just described—PTSD integration, personal development, and belief system development. They begin in a narrow perspective, tight to the center (the trauma, the ego, the tribe) —initially seeing only the PTSD wound, the toddler's "Mine!" the fundamentalist's "The only truth!"—and advance to broader circles, taking in more people, more information, and embracing wider and wider views.

Growth always brings expansion. This is why, in building our retirement home, we put the living areas on the top floor with the best view of the marsh, looking at more of the surrounding landscape, seeing farther with each increase in height. With stages, the ideas and worldviews open up to take in more. The stages all move, they broaden, they all change, and they all move in the same direction of opening wider! With babies and ideas and consciousness itself, nobody taught them how to do it; it's innate, it just naturally happens.

That built-in intelligence just gets better. For me, this is where the exciting part comes in, because my excitement is about mathematics. How can we see that the changes in our thinking are not dangerous to spiritual thinking, are not a satanic attack? What is natural in the experiential world? Enter the mathematics of fractals, cleverly disguised as cauliflower and broccoli—and yes, this will make sense in a moment.

FRACTALS

What we have found in this seemingly random searching is a series of remarkable parallels between the stages of individual development, measured in months and years, and then the stages of development within the long history of consciousness itself, measured in millennia. The pattern is identical. Further, an ideology or religion will exhibit similar changes in points of view, measured mostly in decades and centuries. The time periods change but the behavioral sequence holds across scales; it seems to be a universal pattern. We have also found disruption (chaos) to be an essential aspect of Being. Surprise has been an element at every turn. Mathematics says so!

Euclid meets nature

The geometry we learned (or didn't learn) in school was classical geometry, coming out of Greek philosophy and dating back more than two thousand years. It was all about shapes and space, with clean, smooth, straight lines described by consistent, precise formulas. It is logical and built around integers, whole numbers.

What almost none of us noticed was that nothing in nature exhibits the straight, perfect lines of classical geometry. What the questioning mathematicians noticed was that whereas Euclidian geometry (the classical) is based on integers, whole numbers, nature tends to use fragments of integers, or fractions. It was impossible to describe in a system; the forms of nature were too complex to compute and deal with by hand. And then came the public advent of computers in the 1980s (that time period again!). In the years when computers were beginning to make a move into business use, an IBM engineer named Benoit Mandelbrot was working on these questions of natural forms and called his mathematics "fractal," from the Latin root of "fracture," and "fragment" and, yes, "fraction.' Instead of operations using formulas and theorems, he discovered that fractal math relies on iteration, repetitions, which is the strength of computers.

Fractal, says an online dictionary, is "a curve or geometric figure, each part of which has the same statistical character as the whole. Fractals are useful in modeling structures in which similar patterns recur at progressively smaller scales, such as ferns, tree structures, snowflakes or eroded coastlines, and in describing partly random or chaotic phenomena such as crystal growth, fluid turbulence, the flights of birds, and galaxy formation."[iii] Anything with a rhythm or pattern is possibly fractal-like. Wireless cell phone antennas use a fractal pattern to pick up signals and pick up a wider range of signals, rather than a simple antenna.

The pattern is simple: a fractal is a geometric object that is similar to itself on all scales. If you are mapping a section of coastline and zoom in on it, a small segment will look similar to or exactly like the original section. This property is called self-similarity. An example of a self-similar object is the triangle here, called the Sierpenski triangle.

A fractal line's length will always be indeterminate or infinite, because some details at the smallest scale will always be finer than the ruler can measure; it's like the photos of mirror images disappearing into the distance. The familiar concept of dimension applies to the objects of classical, or Euclidean, geometry, not to fractals.

A whole head of broccoli has the same structure as its very smallest floret, and all the branches in between. Cauliflower does this, too. The structure, texture and general shape of each piece, regardless of scale, is the same throughout the whole head. This is exactly what a fractal design does, it replicates its basic self.

The shapes that come out of fractal geometry look like nature. As we all know, there are no perfect circles in nature and no perfect squares. Not only that, but when you look at trees or mountains or river systems they don't resemble any shapes one is used to in maths. This is an amazing fact that is hard to ignore. However with simple formulas iterated multiple times, fractal geometry can model these natural phenomena with alarming accuracy. If you can use simple maths to make things look like the world, you know you're onto a winner. Fractal geometry does this with ease. [iv]

Fractals in action

Now apply broccoli to the human race. The principles work conceptually as well as visually. The smallest fractal element of the human race is an individual human (the tiniest floret of broccoli), next would be a family, then a community or tribe, then a region, a nation, the population of a hemisphere, and finally the largest human fractal element, the whole human race, to which all are connected. The important thing to remember is that *no matter the scale, each human group has the same psychological structure.* [v] (Emphasis added)

Look at the parallels between personal growth and system growth. Both begin with the smallest, tightest unit in its earliest and most unitive state. In the case of personal development, this is any human infant experiencing only itself and its needs, a pre-verbal, helpless, non-participatory, probably adorable, pre-expressive ego. It knows no separation from its environment and is provided with every life-sustaining need.

To see the parallel with the earliest system stage, go back to the infancy of the human species, sometime around 200,000 years ago during the Paleolithic age. The human species was sparse, living in small family groups which would later become tribes, their basic skills allowing the creation of the first stone tools and art, their survival needs provided by Mother Earth, who gave food through their hunting and gathering, as well as water and shelter. The earliest stories were drawn on cave walls. In terms of humans as a species, this was the infancy developmental stage, brain evolution with limited cognitive intellectual development and an infancy consciousness. They knew no abstract separation from their environment, which provided every basic need. Protection came from magic.

Many, many generations passed, until here and there a creative thinker—or possibly someone tired of all that walking—thought, wouldn't it be more convenient if the plants were closer. And of course, here came the first stationary crops around ten thousand years ago, needing stay-at-home people to till and protect them, all according to the plants' schedule, not the humans'. Infancy was behind them now. Life had become by necessity more tightly organized. and more labor-intensive, but always with time for stories.

By four thousand years ago, with more stable nutrition, life spans and health and populations increased, and there were quarrels over good land and tribal status, and power. From time to time around the globe, populations were on the move, fleeing drought, famine, each other, looking for land, for peace. Favorite stories appeared on pieces of clay. Elders and grandparents told stories of a fabled olden time—the Dreaming in Australia, the Days of the Ancestors, before the troubles began, when no one had to work so hard, and humans could talk with animals. In the Middle East, it was said there had been a fabulous garden called Paradise, where a god could walk with them, but then a fall.

As Jean Gebser's monumental work shows, human consciousness itself, at its enormous scale, was following a developmental trajectory paralleling the one we see in ourselves and in our own children and grandchildren. It is not all at the same time but the same template. Following this conception to our own time, it is possible to see where we are. In infancy in the Pleistocene and having globally gone through puberty in the 1500's, we now have the consciousness structure of late adolescence, which is characterized by the individuation process and a fixation on separateness and achievement. Sure looks that way to me!

This is what I have been looking for: "How does this place work?" Not man-made, logical, synthetic, intellectual arguments, but nature. The real stuff: that our consciousness follows a natural template, that our beliefs lean toward that template, that the cognition we bring to an NDE is following natural patterns of growth and development. That is close enough to a sense of Being for me to see a trace of Creation. It will do. I can say my thanks.

FORMS

All along, we have been talking about forms.

And then there is this, from Diogo Valadas Ponte and Lothar Schäfer: The article is "Carl Gustav Jung, quantum physics and the spiritual mind: a mystical vision of the twenty-first century," published in PubMed. Diego Valadas Ponte, a neuropsychologist, is General Director of the Portuguese National Stroke Association; the late Lothar Schäfer, a quantum chemist, was Distinguished Professor emeritus of physical chemistry at the University of Arkansas. What advanced mathematicians make of their work is beyond my powers of discernment, but I cannot recommend it highly enough for any layperson who is struggling to understand quantum basics.

Not only does today's physics speak of illusion, but it speaks in mathematics (that issue again). What the late Renaissance world lost was a cosmology. What we lost was substance and interdisciplinary communication. But there are exceptions, like the one following. At the time this next quote was written, one author was pursuing his PhD in psychology (and is today a neuropsychologist) and his co-author was a chemistry professor at the University of Arkansas. I say, as an ex-teacher, that you can always tell the really good teachers by the clarity of their exposition. This one article did more to light my path through quantum theory than the stacks of books I've tried over the decades.

The first aspect of the quantum world that we have to consider concerns the fact that the basis of material things is not material...

Electrons are tiny elementary particles: they have a definite mass and, whenever we see one, it appears as a tiny dot: for example, as a flash on a TV screen or a little mark on a photographic film.

In contrast to their appearances, the electrons in atoms and molecules aren't tiny material particles or little balls, which run around the atomic nuclei like planets around the sun, but they are standing waves: when an electron enters an atom, it ceases to be a material particle and becomes a wave. (Is this what happens when we die? Another pattern!) We owe Max Born for the discovery that the nature of these waves is that of probability waves...

By the way in which it describes the world, quantum physics has taken science into the center of ancient spiritual teachings. For example, molecular wave functions have no units of matter or energy. They are pure, non-material forms. The same is true for Jung's archetypes: like the wave functions of quantum systems, they are pure, non-material forms...The discovery of a realm of non-material forms, which exist in the physical reality as the basis of the visible world, makes it possible to accept the view that the archetypes are truly existing, real forms, which can appear in our mind out of a cosmic realm, in which they are stored. Thus, we can confirm here, on the basis of the quantum phenomena, Jung's view that "it is not only possible but fairly probable, even, that psyche and matter are two different aspects of one and the same thing"

"The aspects of the psyche that [Jung] discovered are so profound, that they go beyond the limited concerns of the human psyche, making it possible to think, for example, that the universe itself is conscious and our own consciousness is connected with the cosmic consciousness

By studying the human psyche, Jung discovered mental properties of the universe, which Classical physics had suppressed: Quantum physics has now brought them back.

The facts show us that there is a non-empirical realm of reality, that doesn't consist of things, but of forms. These forms are real, even though they are invisible, because they have the potential to appear in the empirical world and act in it. They can do this in two ways: they can find consciousness as thoughts in our mind; and actualize as material structures in the external world. Thus, the conscious and empirical world is an emanation out of a realm of mind-like forms, and quantum physics is a form of psychology, the psychology of the cosmic mind. In the same way Jung's psychology is also a branch of physics; that is, the physics of the mental order of the universe.

A THREAD ON THE PARANORMAL AND SYNCHRONICITY

In the course of my eighteen-plus years of education, since second grade there has never been a math class I could not (and did not) flunk at least once. Math demands a way of seeing which I almost absolutely lack. (Had it been taught as metaphor, I might have done better.) However deep my blindness, mathematics as a language goes into areas so absolutely essential to the deepest questions, it must be addressed, even by me.

As a laywoman, I work with what's at hand, which is why the Universe has granted me mentors on Chrome and metaphors on keyboards, and, in my senior year of university, a math professor whose about-to-be-failed course was threatening to keep me from my BA degree. It was my third time through the required course, and his equations were still a foreign language. The prof, may his memory be blessed, agreed to my volunteering to write a paper, in words (my strength), demonstrating some understanding of a mathematical topic. I wrote five pages on Einstein's theory of relativity, and the prof passed me with a D. I am honoring that personal tradition in this chapter. I cannot speak about fractals in their native mathematics; but what they say with numbers has fascinating things to say in words also.

Here's what happened in the fractal thread, with more synchronicity.

One day in the early '80s when several IANDS Board members were in the office, a question came up about the historicity of NDEs. When I asked, "Isn't there something in the Bible about this kind of experience?" the response ranged from humor to outright scorn. Why would the Bible have anything to say about NDEs? I was not convinced by their skepticism, but this was before home computers, and I did not know where in my Bible to look and was without a church at the time. The question would have to wait.

Not long after, a social worker friend called to ask a favor. She was to speak at a wealthy church about local poverty and wanted some support. The church was so big and so rich, I had sworn never to attend a service there, but she wanted help, and I was heading a social services training program at the time, so of course I would go with her. We would attend worship and go to her forum afterward. And that was the Sunday the big, wealthy church announced an upcoming two-year Bible study course. The course had only two requirements: participants had to agree to teach the course after completing it, and it was for members only. So I had to *join* the big, wealthy church to get the answer to the NDE question. And I discovered that indeed, there are plural experiences like NDEs in the Bible. And I had to teach the course, which I did twice, which introduced me to enough people that it was the women's association of that church which six years later provided tuition for my Master's degree at a nearby Roman Catholic college so I could learn about mysticism.

Threads. Always, threads. Butterfly effects. Things begin to hold together—the paranormal, quantum reality, depth psychology, mysticism, religion and non-religion, anthropology, consciousness. It's a lot to take in. Pay attention, because Something Is Going On.

When teaching that first Bible course, I noticed vaguely that one could talk about the Bible and spiritual growth and development in autobiographical chunks—time of innocence, time of chaos, time of wilderness. That was cool, but it didn't help explain my NDE, though it did persuade me to pay closer attention to the biblical psychology. That got me to the reams of study about hell, and to my fascination with stages. I was still no closer to any revelation about being an illusion.

This past year I had occasion to make the six-hour drive to visit my younger daughter. The first time I was in her church, a very cool hospice chaplain introduced me to a book titled *Integral Christianity* about stage development in the history of Christianity, based in part on the thinking of Ken Wilber, the American philosopher and scholar best known for his Integral Theory. Integral Theory is all about patterning and suggests a synthesis of all human knowledge and experience.

As mentioned elsewhere, I have been wondering for ten years how to approach writing this book, and the *Integral Christianity* book kicked my mind into gear. A few months later, the pandemic arrived and the world went into what has amounted almost to house arrest (for those of us who take it seriously). There I was, with unfettered time to work on this book, worried by a sense that there was something I had been missing, and here came a book about stages in terms of religion and denominational development; and the stages for religion were very much like stages for individuals. Then, on a whim, I began to re-read Ken Wilber's *Up from Eden*, which steered me to Jean Gebser's pointing to stages in the development of consciousness itself. That led me to line up the three developmental stage materials together, and when the same pattern jumped out of their different scales, I wondered, fractals? Could they be...?

This, let me be clear, is the way the paranormal often works: one tiny step at a time, noticing *how* things were happening. Skeptics will dismiss these meaningful happenings as "just coincidences." With a synchronicity, it feels as if coincidences have been steered.

So there I was, wondering if the similarities I was seeing might have anything to do with fractals. Googling immediately brought an article, "The Fractal Nature of Human Consciousness," by California businessman Carl Zdenek, with an informed, sophisticated, and somewhat different take on the developmental patterns which had so engaged the author of *Integral Christianity*. The Zdenek article included information about *attractors*, the term for invisible form-making forces in the language of fractal mathematics. Once again I wondered, "That sounds a lot like archetypes. Could it be?" And when I Googled again, the first title to jump up was a lengthy article by Jungian analyst John R. Van Eeenwyk, the stunning "Archetypes: The Strange Attractors of the Psyche." Bingo!

That was followed within days by my discovery of the superb Ponte and Schaefer article, "Carl Gustav Jung, quantum physics and the spiritual mind: a mystical vision of the twenty-first century." That single article is so lucidly written, suddenly many blurry questions came into focus. Suddenly I knew where I was going with this integration. There are other threads to weave in, but this is the heart of my revelations.

So here we are—Fowler, Wilber, Gebser, Smith, Zdenek, van Eenwyk, Ponte and Schaefer, and me—theologian, philosopher, philosopher, clergy, businessman, Jungian analyst, neuropsychologist, and quantum chemist, and me–none of us is a professional mathematician, we have never met in person and some of us have died; yet here we are together in this discovery of how fractal geometry and chaos theory speak to our lives, and possibly to near-death and other experiencers elsewhere. What we are as a group, I think, is searchers for meaning, for links holding the world together, for patterns that hint at purpose and intent.

I mentioned that the route of my integration is on display in the topics of this book's table of contents. I also mentioned threads. This is what I mean. Skeptics will say coincidences are purely random and meaningless. I say from my own experience of life and synchronicities, it is as the threads gather and become long enough to braid, that meanings emerge and transformation begins.

We cannot afford—and the experiencers cannot afford—to wait 350 years to realize we have been thinking backwards rather than in the now.

THE STAR TREK GOD

.According to legend, and biblical record, and Homer—and, considerably more recently, Julian James in *The Origin of Consciousness in the Breakdown of the Bicameral Mind*—heroic individuals of ancient times commonly got to be that way because they were following the commands of the voice of a god speaking to them directly. They actually, physically, *heard* the voice speaking. These stories poke like arrowheads out of the *Iliad*, the *Odyssey*, and Greek myths. We find them in the oldest sections of the Bible and in other stories that date from the Bronze Age and perhaps even earlier. So, for example, Abraham, prospering in the city of Ur, hears the voice of God telling him to pack up his considerable household and walk 1,400 miles to a place he doesn't even know; and Abraham becomes the Patriarch of Judaism, Islam, and Christianity because he promptly obliges.

In those olden times, heroes and patriarchs heard, obeyed, acted, and then explained to everybody else what was going on, and the people paid at least some degree of attention because, although they might be amazed, *they weren't put off by that sort of revelation.* They didn't think it was impossible. In time, those individual incidents became the founding stories of holy places and temples; in the monotheistic traditions, they became the ancient heart of the Bible and the Qur'an. Whatever form the experiences took, the messages have come down through thousands of years:

> *The Sacred is here. Pay attention. This is how you are to act. Care for each other and for the world. Be just and merciful.*

Eventually, fewer people heard the Voice directly, but some did, and became prophets. They had remarkable visions or understood the meaning of dreams, experiences in which they recognized a mandate from the sacred:

> *The Sacred is in your midst. Pay attention. This is how you are to act. Care for each other and for the world. Be just and merciful.*

What happens today to a person who hears the voice of God? We have no cultural acceptance for such people, or for wandering prophets; we refer them for psychiatric care. The prophets inhabit our homeless shelters. In this high-tech age, we are comfortable with journalism, not poetry or revelation. We give to science the place once held by religion as the doorway to ultimate truth. How is God to talk?

Here comes the wonderfully comical, paradoxical, mysterious (and hilarious) way of the universal sacred! Just as the aliens in Star Trek always spoke English so we could understand them, the Most High at work in history—what I have called the Star Trek God—does the same.

In a scientifically-minded, secular, and skeptical culture which has been quick to dismiss religion as outmoded, there is still one fragment of personal reality in which almost everyone will still cautiously acknowledge that "funny stuff" can happen. That is the fragment associated with death. And in 1975 a young physician named Raymond Moody published a small book about accounts he had heard, stories of what he called "near-death experiences." And because he had a scientific education and made no alarming claims about the stories, his book was not only socially acceptable but downright fascinating, and the media picked up the stories.

It is a further curiosity that for many of those people whose accounts have been reported, the near-death experience occurred, like those of old, in places of awe and reverence, where they were attended by the esoteric rituals of high priests. The difference is that today's priests are most likely physicians, and our temples carry names like University Hospital.

It is also worth remembering that the prophets of the past were not special; they were ordinary people going about the business of their lives until they were interrupted by transcendence. Their lives were just as disrupted by their experience as are the lives of today's near-death experiencers, and people wrote books about how to cope.

There are more NDErs than there were prophets. And so, a couple of thousand years later, this very different, scientific culture has been able to hear the stories of tens of thousands of Miriams and Jeromes and Isaiahs: the ordinary people who have had NDEs and believe they have learned something true in them. They give us the message of their experiences:

> *The Sacred is in your midst. Pay attention. This is how you are to act—love, serve. Be just and merciful.*

The underlying message has not changed since the Bronze Age. The Sacred always speaks in whatever way we can understand. And¾all those long, solemn philosophies later, and after all the learned debate or scoffing¾how can one not burst with delight at the sheer humor of it?

Baskets of Meanings

I think of this model, then, as the "Star Trek God," and I believe it is true, although I do not believe it to be literal fact. Like all myths, it is true because it is a way of describing a pattern, making a workable model of something too big to grasp directly.

In this sense, all religious stories are like baskets in which we carry mysteries so they can be transmitted across generations. Over time, as history unfolds, understandings and interpretations shift. Stories originally told around Bronze and Iron Age hearths move into the Middle Ages, and on into the Age of Enlightenment, the Gilded Age, and now the Nuclear Age and beyond. The same stories are heard by different ears in different times and understood from vastly different world views.

It's not the fault of the ancient stories themselves that we, today, try to read them like newspaper accounts; it's not a weakness in the narrative itself that leads us to scoff when we find a plot line factually unbelievable. The weakness is our own, because we insist that the stories be something other than what they are. Deep stories lead double lives: one life on the surface and another moving in the shadowy underneath. These stories are, in my metaphor, 'baskets of meaning,' carrying simple narratives that point to a deeper reality *beneath* the stories themselves. And so this story basket continues, moving toward deeper meaning:

Dancing in the Dark

With the debut of Science three centuries ago, Religion lost first place on the dance card of Western civilization. The newcomer was dazzling and accomplished and had soon whisked half the world out of agriculture and into the Industrial Revolution and on toward a technological wonderland. Between then and now, Religion has still found a few partners, but Science snapped up the sexy ones (and, of course, the funding); so after a while R. was sitting glumly on the sidelines with her confidants, lamenting that no one knew how to waltz any more. "Dance," she said, "is dead." Late-nineteenth-century physicists said the same thing about their discipline, in the confident assumption that physics had learned all there was to learn about matter. Both pronouncements were premature.

Within a generation, quantum mechanics had produced the astounding discovery that an atom could be subdivided, and what was more, that when looked at from one perspective, electrons behaved like solid particles; viewed from another perspective, they seemed like electromagnetic waves. Particle and wave; matter and energy. With this came the unsettling certitude that matter—the good, solid, physical, dependably measurable *stuff*—was dizzyingly otherwise: not *solid things* so much as *fields* of fizzing, unpredictable energy and infinitesimal entities at enormous distances from each other.

Consider the implications for yourself: Sub-atomic particles make up atoms; atoms make up molecules; molecules make up cells; cells make up organs; and organs make up *us*... which leads to the incontrovertible conclusion that our very own bodies which seem so—well, *physical*, are constituted overwhelmingly of space, occasionally interspersed with bits of dynamic buzzy stuff. But if we're not solid *stuff*—if we're mostly space—what are we? Who are we? Where do I stop and you start? Where are our edges?

Philosopher/psychologist Jean Houston says, "We all have leaky margins." (Think of your feelings of discomfort in a crowded elevator where everyone is 'leaking' on everyone else.) We are a whisk of buzzy atoms and sub-atomic particles, all in a whirl and dance. We are a soup of photons. British physicist David Bohm says, "Matter is frozen light." Look at yourself, and at the people around you—frozen light. You are the salt of the earth. You are the city that is set on a hill. You are frozen light. You are particle and wave; matter and energy¾body and spirit…like electrons, and archetypes, and imaginal events.

And the Star Trek God laughs.

MAPS, FORMS, AND TERRITORY

The universe is our elephant, and like the men of the old story, we try to describe that part of it we grasp. Although Universe can neither be taken in by a human mind nor described completely from any one perspective, that has not kept us from trying: Bronze Age nomad, twelfth century monastic, 18th century mathematician, 20th century physicist, 21st century author or file clerk—we all look out on the same universe, look to the limit of our senses, and interpret our description. These models have developed into theories, doctrines, belief systems: all the forms of religion, the varieties of philosophy, types of spiritual discipline, branches of science, political parties, each believing itself to be the truth about the elephant of the universe.

Is this view of mine the end of faith? If beliefs, if theologies, if doctrines merely describe models rather than an underlying stability, what can we trust to be true?

My experience has been that this understanding provides the beginning of faith, a bedrock of unshakeable belief. What struck me some years back was the stunning realization that, as different ranging as people's belief models are, many display a remarkable underlying consistency. This is known as the "perennial philosophy," the core of all spiritual systems. We can trace their common characteristics in the models described by Scripture, by the great spiritual technicians called mystics, by near-death experiencers, and by today's particle physicists. Each set of descriptions is an attempt to picture Reality . . . Creation . . . the Whole. Each is a map of the great cosmic territory. One model is made up of themes shared by sacred writings. These traditions acknowledge a "something other" in creation, a sensed holy presence, a powerful creating and shaping force; in English, it is often called "God." These scriptures of religious tradition include abundant mention of radiant light. They speak of the importance of caring, of loving one's neighbor, even one's enemy. And whether in stories of Eden or of lost Ancient Ones, they tell of an unimaginably distant time of harmony and wholeness.

A second model is present in the writings of the world's great mystics, past and present who live with a radical sense of inbreaking spiritual experience. Their lives have been captivated by a sacred Presence, an unseen Beloved. They describe their experience in terms of light and darkness. Overwhelmed by the centrality of unconditional love, they attempt to describe their flashing moments of union with the sacred Presence as a seamless and ineffable whole.

The third model comes from the accounts of near-death experiences across the centuries. People who have had a near-death experience talk about light¾its positive presence or absolute absence. They tell of presences, sometimes of a sacred Presence they may call it by a name familiar to them from their religious tradition or refer to simply as a Being of Light. Their lives afterward are driven by the conviction that the most important thing is love, that everything is connected, that there is a oneness to everything in the universe.

From a fourth and quite different realm come the models of quantumh physicists, described not so much in words as in mathematics. They have shown that the most basic bit of existence is the quantum, the smallest bit of something, which includes a photon, the smallest particle of light, Increasingly they say there seems to be, somehow, a shaping intelligence behind (or within) the workings of everything that is. They speak of their search for a unified field theory, following hints of a theoretical commonality linking all things. And particle physicists have documented the unarguable truth that at the sub-atomic level, there can be no objective observer and observed, for in a mysterious dance of oneness everything at the quantum level interconnects and interacts.

---

[i] https://fractalfoundation.org/resources/what-is-chaos-theory.
[ii] https://reylab.bidmc.harvard.edu/pubs/1990/sa-1990-262-42.pdf.
[iii] Dictionary Https://Www.Google.Com/Search?Q= Fractal.
[iv] Dallas, George M. Https://Georgemdallas.Wordpress.Com /2014/05/02/What-Are-Fractals-and-Why-Should-I-Care.
[v] Vaneenwyk Https://Www.Jungiananalysts.Org.Uk/ Wp-Content/ Uploads/2016/10/Van-Eenwyk-J.-Archetypes-The- Strange-Attractors-of-the-Psyche.pdf.

PART III.

WHAT WE BRAID TOGETHER

CHAPTER 9.

INTEGRATION: MARINATE IN NDES

In March of 1982, on my first day of work at IANDS, Ken Ring said to me, "Here are the files, and there are the bookcases. Start reading. Don't do anything else for three weeks."

It was such a heady privilegle, being there right at the beginning, when the field of near-death studies was brand new and there was so much excitement and shininess, big-time television executives calling, books hitting best-seller lists, the phone ringing, a world agog to hear what might be actual heavenly experience accounts. Even now, without all that swoop and bustle, if you immerse yourself in the classic, beneficent NDEs—the ones about love and reassurance—your life will change.

Raymond Moody and the early researchers—Ken Ring, Bruce Greyson, Mike Sabom, and Margot Grey—realized quite soon that it was not only near-death experiencers who reported changes in their lives. The researchers also found themselves losing some fear of death, being more open to a loving disposition, feeling more trusting of life. And then the same shifts were being reported by people who had been reading the books about the peaceful, sometimes radiant NDEs. The experience effects were contagious!

For establishing a foundation of trust, I cannot recommend any healing activity more strongly than a period of saturation–marination, I call it–in the beautiful NDEs. Especially for anyone with a deep fear of death or who has had a troubling (or worse) spiritual experience, immerse yourself in the first-person accounts of peaceful and loving NDEs and similar STEs. Yes, I had to work through a stage of "How come they got those and I got *this one*," but the reading was building a foundation for integration. However, all these years later, I add one caution: Read the accounts, the first-person descriptions of the experiences themselves and do not at first get caught up in explanations or interpretations. Simply let the experiences speak for themselves..

Today there are differing schools of thought about the source and meaning of these experiences, and arguments about how they work; there are even secular fundamentalisms about spiritual experience. People will explain endlessly how life before birth operates, and how reincarnation or past lives are structured, and how heaven is organized; they will explain *why* experiences occurred to particular people. Those metaphysical details are not the experience; they are add-ons *to the experience*, which may or may not be accurate but which are deeply believed by some.

They are not the experience; they are intellectual explanations *about* the experience. If you are interested, by all means follow them up later, but at first, all of that is pointlessly intellectual and a divergence from what I am recommending, which is that you saturate yourself in the lovely experience accounts until you discover a level of trust in the universe. Whatever else we encounter in the universe, these experiences are common. I want to encourage you simply to read or listen to first-person accounts. IANDS has a large collection available online, as does NDERF.

The fact is, *there is no empirical evidence* to say why people get NDEs, or which people, or what kind of experience. *There is no clear-cut empirical evidence about any of the 'whys'* except that these altered state experiences happen. *They* are the experiential reality. Argumentation is not the point. The point is: *These affirming, beautiful experiences exist and that can help us learn to trust the universe/God/whatever.*

For the sake of this healing exercise, avoid the theorizing *about* NDEs and how they work. Simply dive into the first-person accounts. Listen, observe, let them soak in, notice which ones feel comfortable and which not. Do not jump onto any interpretive bandwagon. Just listen. This is a trust-building exercise.

I don't suggest staying entirely away from distressing NDEs—they exist, and we need information about them—but going there demands engaging your intellect, which is not what I am wanting to go first. First build your sense of trust in the light-filled accounts. *Do not immerse* in the grim ones; they are so powerful we need to take them in little bits at a time, with more than hell as an interpretive framework. There is time for that; but first, this.

Immersion in the beneficent NDEs will build your foundation. Whatever they are, these experiences exist in their millions. Simply rest in them.

THE MYSTERY OF TRANSFORMATION

From psychotherapist Alex Lukeman comes the heart of a new paradigm:

"When the ego encounters the underlying dynamics of the numinous [the sacred], there is ...the accompanying destruction of traditional and habitual patterns of perception and understanding, including religious belief structures and socially accepted concepts of the nature of human existence and behavior."

When the Western ego discovers it has no substance...Bam and gone! The personal reality implodes. Everything destroyed. But notice:

Lukeman does not say the person's experience has been in the slums, or that the spiritual level is too deficient to get the good stuff. He says ego has just met the sacred, which, as Rudolph Otto pointed out, is *often perceived as terrible.* The devastating upheaval is not only normal but *expected in deep spiritual experiences.*

A difficult experience is not inherently negative, any more than a newborn intensive care unit is negative for its mix of tragedies and rescues. A NICU nurse in Connecticut some years back was regularly booked as a stand-up comic at pediatric nursing gatherings; she had a great line: "Nowhere is it written that life is serious, though it is often hard and even terrible."

The nurses knew just what she meant. These professionals understand the reality of the dark emotions alongside miracles. I borrow and rephrase the line now: Distressing NDEs and the difficult aftermaths of beautiful NDEs are not negative, though they are often hard and even terrible.

In an exchange of emails, I had written to Lukeman, "Can we please put an end to thinking that only mean and unspiritual people have distressing NDEs!" His immediate response was:

> [It seems] a good possibility that a person's NDE experience could be related to psychological and unconscious states, even though outer life experience might not coincide with that experience. A serial murderer could possibly have a joyous experience, a decent and spiritual person a frightening one. Since the unconscious is by definition unknown, literally anything could be lurking there, dark or light, including all of

the collective thoughts built up since humans appeared about death and what comes after. That would be Jung's "collective unconscious" which is quite real as far as I can tell.[1]

Exactly! Sometimes we simply have to get out from under all the religious and metaphysical explanations and clear the decks for long enough to see another perspective. Lukeman points to the hemispheric distinctions in his masterful *What Your Dreams Can Teach You*, saying:

> In the East, the negative is just a different face of God, part of an eternal cycle of change and transformation… The spiritual challenge is to recognize and appreciate these different faces of God made manifest in human form… In the West, we are taught to deny the negative and focus diligently on the "good." Humanity [must] redeem itself through particular actions, contrition and submission as defined by each of the three great religions. Failure to do so will lead inevitably to damnation and eternal torment…Leaving aside theology and opinions or beliefs about who is right, it makes sense that our dreaming mind will choose the familiar symbols of its cultural heritage to represent mythic forces of divine and supernatural good and evil. If we are entering a crisis in our spiritual development, these symbols may appear…These symbols may also appear during times of fear or inspiration in our inner and outer lives.[i]

When, in situations such as mine, the images which appear are *not* familiar, the absence of suggestion and meaning creates great confusion and alarm.

As a near-death experiencer himself, Lukeman's work about nightmares applies also to these nadir NDEs. Here, for instance, are seven suggestions for finding relief from the fear of a terrifying dream or spiritual experience, with a comment from me on each of them:[ii]

> "Realize the [experience] didn't occur just to scare you but has meaning and purpose. This is absolutely fundamental.
> [NEB: I know, what is the meaning and purpose of *that*! The way the psyche works does not always make sense to us, but we are

not the ones in charge. Don't take your first impression as the only meaning. Study up. Do the work the psyche is demanding!]

"You have an innate ability to understand and gain relief from the [experience].
[NEB: But first you have to stop panicking. Breathe. Breathe some more. Keep reminding yourself you have an innate ability to do this.]

"Don't be afraid to look at horrifying images, or think that bad images mean bad things about you.
[NEB: Start with just quick peeks. Build up your ability to stay with them until you can actually look. Saints have also had to deal with these kinds of images; the images do not mean you are bad! It means you are being challenged to deal with tough stuff.]

"Learn to step away and consider the image objectively, without emotion, if possible.
[NEB: Sounds impossible. You can learn to do it. And it's great experience for all kinds of situations.]

"If you get an idea of the meaning, you get two good results: You won't have to have the dream again and you have practical advice for your real, outer life.
[NEB: True!]

"Remind yourself that nightmares can open the door to love and healing in both a psychological and a physical sense. They are a gift from your unconscious mind, even though they seem frightening when they occur.
[NEB: This is the deepest shift of understanding. This *thing* is a gift. Our job is to figure out how it is true.]

"Let your mind relax; freely associate images, feelings, and memories."
[NEB: This is how the psyche works. See what arrives. Get help if things get too much to handle alone.]

THE RELIGION QUESTION

The plain fact is, in contradiction of what we expect, most altered state experiences do not manifest in explicit terms of any specific religion, although experiencers who encounter familiar images interpret them according to their own background. Only rarely do the elements of altered state events have to do with formal theological doctrines, dogmas, or laws.

Where NDEs touch specific religion is in experiencers' interpretations. For instance, say a person having this type of experience encounters a compassionate spiritual entity. That is a common occurrence, and, being human, we immediately want to name it. Who is that figure? The compassionate spiritual entity appears as who the person needs/expects/will listen to, most commonly a relative or a religious figure. When an entity is identified in the Western hemisphere as a religious figure, it is usually Jesus or Krishna (though I'm fond of the story of the woman who said the welcoming figure was St. Jude, because "I pray to him all the time, and he knows me best."). Finding an image they understand as familiar, the experiencers tend to return to the world full of religious conviction and gratitude. Maybe the entity really is Jesus, or Krishna, or Moses, or Elvis; they do not wear nametags, and the point is what the experiencer believes. Not all NDEs, however, present such specific images.

An absence of strong religious content is interpreted in a variety of ways. For those with fundamentalist beliefs, a blissful NDE may likely be considered satanic if it does not stress wrath and judgment; on the other hand, a dreadful, violent NDE supports the doctrine of hell. In more liberal or mainstream religious systems, the power of unconditional love in the Light will be understood as the presence of God or an affirmation of human worth. When the experience is difficult or horrific, a common response is fear and overwhelming guilt; a good many of these eperiencers search for a religious community with strong rules of conduct and claiming the authority to guarantee a road to salvation.

Yet these experiences are so incontrovertibly individual, our responses may be all over the lot. In my case, I had, in effect, been thrust unceremoniously and with no explanation into an essentially Buddhist philosophical scenario and came back in part terrified but also in a towering rage at God and the church for not having prepared me for…*that.* West meeting East is sometimes more difficult than trying Yoga.

As a result, I remained unchurched for a long while. For several years trips to the supermarket took me past a Congregational church which I had never attended, though it was my denomination. One day I was horrified to see, over the wide front steps, a banner bearing a single huge word: J O Y. I seethed with fury. *Joy? What joy?* How could a pastor tell his people that! How could a pastor mislead them, trick them to believe in joy when they were nothing but an illusion? What did he know?

I was so angry, I made an appointment to tell the pastor so. He was an exceptionally gifted pastor who ultimately became a psychotherapist. No surprise there, because by the end of our first meeting I had agreed to return, not to Sunday service, but to his Tuesday morning mothers group for coffee and conversation about whatever was on our minds. It was ideal. We hardly ever talked about religion.

Although I rarely spoke, and never about the NDE, the coffee hour served as a cushion, a substitute therapy group until we moved. By then, although there was no miraculous cure, I had experienced enough inclusion in an atmosphere of utter acceptance, that the edge of my anger had been softened; I could once again be part of a group.

There is no shazam punch line to this anecdote, only the mention of what a profound difference can be made by some gifted listening and making safe space in which a tormented stranger could begin to settle down and recover herself. This is what the best churches do. If you have never encountered this in a religious group, you need to get out more. The pastor, Carl Christensen, was truly a gift, and there are others like him in all faiths.

Carl suggested books I might like. In the twenty years immediately following the NDE, my reading was shaped by his recommendations as I struggled to find a more satisfactory understanding of religion, and by the requirements of the one or two graduate courses I took every semester. The readings included Piaget, Lawrence Kushner, James Fowler, Carl Jung, Alan Watts, Thomas Merton, Morton Kelsey, and John Sanford. I was being quietly edged toward the transpersonal realms.

Although my church relationship remained decidedly patchy, Carl got me far enough back so that one thing could lead to another and I would eventually, while at IANDS, do a Masters Degree in Pastoral Ministry and Spirituality at a Catholic college so as to learn about mysticism. These paths are not direct!

The degree gave me academic respectability more than religious belief; I did not become Catholic, did not stop questioning, did not abandon faith but remain an outlier in my religion. I like author and theologian Marcus Borg's borrowing the Buddhist story that religion is like a pointing finger: our purpose is not to watch the finger, but to see the moon. Like Borg, I am convinced we have mistaken doctrinal beliefs *about the finger* for the moon itself. NDEs and STEs can be a way of prying us away from unhealthy and uncompassionate attachments so that we can ask different questions and grow closer to the moon. *But then we must not mistake NDEs for the moon!*

We have attached to beliefs as if they were facts. Today, my belief life is broader than most Christian theological language. To belligerent atheists I might say I do not have an "imaginary friend in the sky" but am being attentive to a perceived forcefield. For others, I would borrow Borg's language again and say I am "a nonliteralistic and non-exclusivistic Christian," committed to living my life with God within a

Christian tradition, even as I affirm the validity of all the enduring religious traditions. Borg and most progressive Christians share the view that being Christian is not about believing the Bible, *per se*, as factual or about "believing in Christianity." Rather, being Christian is about deepening one's relationship with the God to whom the Bible actually points (which is *not* the caricature 'angry old man' of so much bad theology). Caring about others and the world around us is a natural outgrowth of that deepening relationship.

That faith remains foundational and still drives my questions: We are not the designers of this world; how does it work? What is the message? What makes sense? My mother's Kansas pragmatism runs strong to this day. I know that at deeper levels 'something is going on' and am looking to understand as much as I can. Where I am is way too abstract, too wordless to be mainstream. An image came to mind some years back, of Church as an oceangoing liner full of passengers, with me in a small boat, rowing in another current but moving in the same general direction, still calling back and forth with a few people on the ship.

EXPLAINING HELL AND DISTRESSING NDES

An unhappy history

I have found it sad to discover how many people's first assumption about a distressing NDE is based on an immediate sensation of falling, or the image of a goat, or a snake, with the assumption that they are headed for a guilt-based external hell with all its Dantesque implications. Almost as many slam the door on any mention of religious belief and call the images hallucinations (and then discover that the images sit just on the other side of that door and do not go away on their own.). Whatever their personal specific religious beliefs and despite space travel and Hubble photographs, residents in Western societies have absorbed the three-tiered cosmology which says God/the Sacred above, Earth in the middle, Hell somewhere below our feet with all its gory imagery. Despite my liberally spiritualized religion, I have always lived in this Western culture and had to struggle through some of those same doctrinal swamps.

An alternative view

With its fires and pits and eternal physical torment, what the West thinks of as hell is not a fiction—because we know that such NDEs exist. However, despite what biblical literalists maintain, the notion of eternal torment as a *place* developed well after the Bible was written. It is an *interpretation* made by human minds and developed over several hundreds of years to fit changing political and religious needs.

I have written about that at some length in *Dancing Past the Dark,* (that chapter is freely available on the blog, dancingpastthedark.com), and scholars have written books of evidence to the same effect. It is possible to be logical about this *factual* discovery and to disconnect from much of the sickness and terror by cognitive work. That's the easy part!

What is far more difficult is to find release from the grip of the archetypal images of the imaginal psyche. It is one thing to disregard a mistaken teaching, but *what was that experience?* Unfortunately, most religious institutions say little if anything helpful.

The images in dreams and these experiences do not necessarily reflect what the individual consciously believes, and in a few (seemingly very few) instances, the images have no obvious link with the native culture; my NDE had no obvious relationship with my beliefs or with New England Christianity; it was more like a bureaucratic report from elsewhere in the universe. I wish I had known earlier about the work of Stanislav Grof.

For readers stuck on the idea of original sin, I strongly recommend Matthew Fox's *Original Blessing*, which has changed many lives. More recently (and fun reading) is Rob Bell's *Love Wins*. If you've been told he's the Anti-Christ, read the book and check him out for yourself.

Working with the primal terrors will take time, and patience, and work. Therapists report that similar deeply ingrained terrors affect clients with post-traumatic stress from other causes, and that many resist doing the hard work required. The price of that resistance is never getting beyond the slavery to terror, which means they go to their eventual deathbed still in its grip.

And I am going to say what I believe is the ultimate truth: If the God you have been worshipping is too full of wrath to allow these healing activities, that is not the real God.

Distressing NDEs—the Shadow of near-death studies.

Jung and depth psychology have given us the concept of Shadow. All our unacceptable parts—our guilts, shame, weaknesses—are buried in Shadow, where we can avoid looking at them. It is our immature ego which operates solely on the pleasure principle, keeping us mired in what we think is self-interest but which is really our fragmentation. But our deeper Self knows that we must directly confront and reintegrate the repressed contents of our unconscious before we can achieve wholeness.

When we banish our existential fears and painful ideas, hoping that will keep us safe, they instead become our monsters. Ironically, this shortsighted attempt at self-preservation turns on us and becomes either psychopathologies or other growth-inhibiting mechanisms. We wind up totally misunderstanding the sacred messages of psychological ordeals and distressing NDEs.

Distressing NDEs are not dreams, but they come from the same imaginal core of our deep unconscious. Jung taught that nightmares may arise as a symptom of failed integration, an unhealthy split of the conscious and unconscious parts of the mind. This is why his approach to nightmares was to encourage the dreamer to accept the frightening elements as parts of themselves: "A persecutory dream always means: This wants to come to me. You would like to split it off, you experience it as something alien — but it just becomes all the more dangerous." [iii] Jung did not mean acting out unconscious energies or surrendering to their control, but rather acknowledging their reality within us and respecting their role in the healthy functioning of our minds.

Hell lives inside us, burning as the fires and torments of our shadow. That is what we meet in a distressing NDE. We have to be brave enough to confront our shadow, our demons, our darkness, and move through it. It is not punishment; it is an invitation to growth, and to wholeness.

Conclusions about judging an NDE

Any NDE is a rite of passage – it is a temporary state. It has a before and an after.

The kind of NDE an individual has is not a permanent measure of their character. Beautiful NDEs happen to dreadfully flawed people, and painful NDEs happen to wonderful people. We have to stop accepting blanket negative judgments about people who have a difficult NDE—sometimes it's just a spiritual bad hair day.

The event does not allow us to go back, we have to go forward. So we have to learn enough bravery to walk into the questions we fear the most. As individuals and as IANDS, we are being called to look deeply at our resistance to the disturbing NDEs.

Rather than looking from the filter of our terror, learn to be primitive enough to see ordeal as a challenge, as a gateway to other realms, as a source of potential warrior pride—of survival and triumph.

There is no global answer. The question is always personal: what is the message of the experience *for this person*? The gift of the hero's journey is not always the same old apple. It is not enough to go on YouTube and say only, Oh, it is so scary…You must demand to know what is its *gift*? What is it telling you about yourself? Ask: What do you want? Why are you coming to me? What is your message? What is your question of me?

Like those people of the late Renaissance hearing that their earth had come unmoored and the cosmology which had been their rock for thousands of years—we have to be brave enough to admit that our comfort zone has to stretch way wider than we are ready for.

What do we get with this approach? We get to let go of the infantile belief that every difficult experience means that we are being punished; we let go of hell. We drop knee-jerk judgments about people who have scary rites of passage. We learn more of truth, which can then be passed around. We learn empathy and compassion. We discover more about the paradigm of the new cosmology which says there are no separations—that every kind of experience is our own. And as we let go of the old patterns, we move further toward our own wholeness.

PRACTICAL STEPS TO RESOLUTION

What can help a person recover from a powerful altered state experience, especially a difficult one, or a difficult phase? My own trek says, just keep going, one step at a time. The activities in this chapter can help. The material here is abbreviated from Chapter 16 in *Dancing Past the Dark*. The suggestions are things I have lived myself and found helpful. They have been read and okayed by numerous therapists, so I am willing to re-publish them.

Initially

In the early days/months/possibly years following a deeply disturbing experience or its effects, it is important to *tread lightly*. The only recourse may be to buy time until panic has subsided enough for cognitive process around the event even to be possible. Hope need not be reasoned to be sometimes most important.

Points to remember, especially in the early stages:

- *No study has ever shown that 'Good people have good experiences and bad people have bad ones.'* In fact, the evidence is that very good people, even saints, can have horrific experiences and people considered very bad can have heavenly ones. Reward and punishment do not work the way we think they should.

- Christopher Bache, basing his statement on clients who have worked through the stages of Grof's perinatal matrix, has emphatically asserted (1994, p. 42) that "a frightening near-death experience is an incomplete near-death experience." Although the overall evidence is not conclusive, reassurance sometimes has more value than data analysis.

- Holy figures in all traditions have endured frightening experiences ¾not only Saint Teresa and other saints, but Krishna, Jesus, shamans. Their experiences did not keep them from being holy, so something must be going on other than judging them as bad or lost.

- The Void is traditionally considered by mystics to be the ultimate experience in spiritual practice. A person terrified or confused by it may be like a first-time skier pushed onto championship jumps.

- The conceptions of 'self' and 'reality' in Western and Eastern faith traditions are very different (and note, we're talking *hemispheres* here, not regions of the U.S.). I keep hearing from other Westerner who, like me, have been traumatized by the Eastern conviction that the physical world and ego are merely illusions. With or without an NDE, arguments based on ego death and illusion can create more anxiety than enlightenment.

- *Defuse the emotional intensity by, literally, getting it out.* Alex Lukeman suggests telling the story out loud three times; if there is no listener, saying it out loud anyway will help defuse the anxiety. From my own experience, this is highly effective. Some people may feel less threatened by writing the experience first, then reading the text out loud. See "Get out of your head," below. Also see Lukeman's seven suggestions in the previous section.

Later

Note: The following suggestions assume that although a person may be troubled and fearful, there is no pathology. When there is mental fragility, find psychotherapeutic help before anything else.

At some undefinable time, it will be time to *begin exploring ways of thinking about the experience*. There is no timetable and no prescribed route. These are a few of many possible paths; they are here because they have been helpful in resolving my own experience.

- If you have had a terrifying NDE, work from the

premise that it is *a different kind of* spiritual *experience.* This means there is value in it, not a categorical casting-out or sign of dysfunction. The focus then shifts from "Why did this happen to me?" to "What meaning can I find in this?"

- If there is an eleventh Commandment, it must be, *"Thou shalt do thy inner work."* It is not possible to dance the spiritual dance while the floor is littered with psychodynamic rubble. Begin to clear the floor. If you can, find a therapist to help you identify your personal issues; if you can't afford long-term help, pick one issue and work with a therapist short-term; if you can't afford that, apply to a clinic; and if you still can't afford that, get a library card and begin reading. This is not a short-term project and it is not the only task, but it is *absolutely essential.*

- Be careful on social media. Among other hazards, once we know how much of the universe is hidden from ordinary view, we may become susceptible to conspiracy thinking. That is a giant, dangerous rabbit hole. We need to learn discernment, not to assume that anything not public is a plot or manipulation.

- If you are dealing with a past-life scenario, find ways to keep a clear distinction between the metaphoric, symbolic meaning of that identity and a literal, factual interpretation of it.

- As soon as possible, which may be quite a long time later, it may be productive to *name the issues* that made the experience so frightening: Was it being out of control; or unfamiliarity? What about that issue makes it feel like such a threat? This provides a focus for your thinking. Some short-term cognitive therapy may be useful, especially with issues related to control.

Discover the techniques developed for dreamwork and imaging. As Alex Lukeman has shown, this information is invaluable. NDEs read like dream material, and decoding symbols applies to deciphering NDEs and similar experiences as well as dreams.

- Keep in mind that no simple list of symbols and meanings is going to be complete or necessarily accurate. The symbols in your experience are yours. A grounded dream course or workshop, or working for a while with a Jungian therapist, may help.

- In myths, the hero often carries a talisman for protection and goes with a companion. Approach your NDE mentally, carrying in your mind (or, for that matter, in your hand) a talisman to protect you. Go with a mental companion¾living or not, whom you actually know or not¾to help you through the experience. Notice what and whom you chose and what significance they have for you; free-associate and amplify what you can say about them. See if events in your experience shift, and how.

- What could you have done differently during the actual experience? Now that you can participate consciously in the events, you are able to change your responses, to observe relationships, to try "going with" the experience. What differences do new approaches make?

- *Expand the symbols; what else could they mean?* Fire, for instance, which we generally interpret as the worst sort of punishment, has an honorable history as a symbol of cleansing, purification, and rebirth (for example, the Phoenix rising). Darkness, weeping, gnashing of teeth may be, not indications of a quasi-physical Hell, but as theologian Hans Kung (1984, p.140) points out, "harsh-sounding metaphors for the menacing possibility that a person may completely miss the meaning of his life." As in dreams, a suggestion of death may point to the end of a life phase or a major change in one's awareness. The Void may feel like ultimate abandonment; it can also represent ultimate unity to someone who is ready to experience it that way. A horned creature, instantly terrifying as Satan, has an ancient history as a guide. The more one knows about symbolic language, the wider the possible range of interpretations.

From Hillsborough Community College in Florida comes this starter list of symbolic archetypes: [iv] Note that they are suggested meanings, not hard and fast rules.

- Light vs. Darkness: Light usually suggests hope, renewal, or intellectual illumination; darkness implies the unknown, ignorance, or despair.

- Fire and Ice: Fire represents knowledge, light, life, and rebirth, while ice, like the desert, represents ignorance, darkness, sterility, and death.

- Nature vs. Mechanistic World: Nature is good while technology is evil.

- Threshold: Gateway to a new world which the hero must enter to change and grow

- The Underworld: A place of death or metaphorically an encounter with the dark side of the self. Entering an underworld is a form of facing a fear of death.

- Haven vs. Wilderness: Places of safety contrast sharply

against a dangerous wilderness. Heroes are often sheltered for a time to regain health and resources

- Water vs. Desert: Because Water is necessary to life and growth, it commonly appears as a birth symbol, as baptism symbolizes a spiritual birth. Rain, rivers, oceans, etc. also function the same way. The Desert suggests the opposite.
- Heaven vs. Hell: Man has traditionally associated parts of the universe not accessible to him with the dwelling places of the primordial forces that govern his world. The skies and mountaintops house his gods, the bowels of the earth contain diabolic forces.
- The Crossroads: A place or time of decision when a realization is made and change or penance results
- The Maze: A puzzling dilemma or great uncertainty, search for the dangerous monster inside of oneself, or a journey into the heart of darkness
- The Castle: A strong place of safety which holds treasure or princess, may be enchanted or bewitched
- The Tower: A strong place of evil, represents the isolation of self The Magic Weapon: The weapon the hero needs in order to complete his quest.
- The Whirlpool: the destructive power of nature or fate.
- Fog: uncertainty.
- Red: blood, sacrifice, passion, disorder
- Green: growth, hope, fertility
- Blue: highly positive, security, tranquility, spiritual purity

- Black: darkness, chaos, mystery, the unknown, death, evil, melancholy
- White: light, purity, innocence, timelessness; also, death, horror, supernatural
- Yellow: enlightenment, wisdom

Use both sides of your brain:

Get into your head.

This cannot be over-stressed: Find the best information you can from as many sources as you can. Find views you agree with and views you don't. Build an educated understanding of the subject.

- Read, read, read. My best help, as always, has been reading, and the best tip I can offer is, when you find an author whose book really speaks to you, check the bibliography to see what the author has been reading, and go read those books yourself.

- Get the facts about these experiences, but also read *outside* the near-death literature. Read the mystics old and new, the history of science, the perennial philosophy, a modern translation of the Bible (not King James, for reasons there is not space to go into here), the Koran and Bhagavad Gita; transpersonal psychology; the 'new physics,' consciousness studies, metaphysics, contemporary theology, general books about religions other than your own; read hymn lyrics and poetry; read the mystics and Joseph Campbell; read dreamwork and imaging techniques, and about healing. If you are a spiritual person, read science; if you love science, read the mystics. (Sometimes the least comfortable ideas may trigger something and become the most useful in the long run.)

- Do not read only near-death, only New Age, only Evangelical, only *anything*. Go for breadth; widen your horizon; find what speaks to you and wonder why.

- Listen to tapes of conferences and lectures. Listen with a critical ear, not to find fault with the presenter but to hear patterns, themes, and inconsistencies. Beware of people who insist they have the definitive answer.

- If you can afford them, go to conferences; take courses. If you cannot afford them, use the library; attend programs at a compatible church or temple (if secular, try Unity, Baha'i, or a Unitarian/Universalist church or your faith's equivalent); find a pertinent short-term adult ed course at the nearest high school or community college. Always, always, be cautious of people who are certain their opinions are absolute Truth.

To experiencers: Through it all, recognize that you are vulnerable to charlatans, bad information, and being ripped off. You will want a guru; be careful choosing your role models. Take your time. Be very wary of conspiracy thinking. Do not commit to any person or group that will not easily let you go.

Remember that spiritual growth cannot be bought; higher price does not equate to higher consciousness. This has been a real issue for me. I simply did not have money enough to attend the workshops I longed for, just as I could not afford the PhD. In the long run, this forced me to recognize my own singular path and keep searching for answers that fit.

Get out of your head.

- Paint your experience. Sculpt it. Weave it. Dance it. If you can't dance, walk it; if you are physically paralyzed, imagine how it would feel to dance it. If you do crafts, make something of it. We are most accessible to sudden insight from our weakest aspect: So, if you ordinarily function from strong verbal skills, try something kinesthetic. If you prefer physical activity, read poetry or keep a journal.

- *Find interpretations that differ from yours.* As Bruce Greyson noted years ago (personal communication, 1985), every interpretation represents a different model of the experience—and they are all *models,* not the real thing. Notice how they are alike, how different. Discover which ones feel most comfortable and uncomfortable, and then examine why —not, 'Why is my interpretation right and that one wrong,' but what do your choices say about what is important to you?

- *Identify the vocabulary that registers with you.* Different models have different vocabularies. Find your vocabulary; for example, some people find that working with psychological terminology opens doors, while others consider it just psychobabble. Then remember that the words build a *model*, not the real thing. No one has *the* single definitive vocabulary or interpretation.

Look for a community

Recognize what can be expected from a resource. A clinical psychologist may be skilled at helping unravel life-history material but not dream symbols. A spiritual director works with a person's relationship to the Sacred, not with deep-seated psychological problems. At any given point, one may be more helpful than another.

Identify a community: The work is always done alone, but others will help. Caution and discernment are crucial; this is harder than finding a dentist.

- Some IANDS support groups are helpful with distressing experiences; others are too fearful, or too enamored of the radiant experience to tolerate the depths.

- Scout the specialized online groups: On Facebook, type 'near death experience' into the search box and click Group. Each group has its own personality; find one that fits you.

- For the person who has given up religion but is still searching, look for a church or temple that better meets your needs today (there's nothing wrong with 'shopping around').

- In some geographic areas, it may be possible to find a spirituality group or book club. Be careful; these can be wonderful but are often the targets of especially needy and/or ungrounded folks who will drain emotional energy and talk nonsense rather than encourage genuine understanding and growth. This same caution applies to some Bible study groups; if they demonstrate unbalanced interests in punishment and predictions of doom, or are absorbed by warnings about satanic influence more than in living and loving affirmatively with the Holy, find another group.

---

[i] Lukeman, Alex. *What Your Dreams Can Teach You.* Audiobook.

[ii] Lukeman, Alex. *Nightmares: How to Make Sense of Your Darkest Dreams.* New York: M.Evans, 2000. 10-14

[iii] C.G.Jung, *Children's Dreams: Notes from the Seminar Given in 1936-1940,* eds. Lorenz Jung and Maria Meyer-Grass, trans. Ernst Falzeder. Princeton, NJ: Princeton University Press, 2008, 372, in *Teaching Jung,* Kelly Bulkeley, Clodagh Weldon Oxford University Press, 2011.

[iv]   Https://Www.Hccfl.Edu/Media/724354/Archetypesfor literary analysis.Pdf

Notes

1. Lukeman, personal communication.

CHAPTER 10.

TEMPTATION, ITS OWN ENDLESS DREAM

by Leigh Gray Kenyon, 2006

First the swirl of madness. That's what it was,
a blizzard of endless choices, endless
possibility, and it had my name.
What could I do? There was no time, no why,
It was raging, and only myself there
to figure myself out. I managed, but
not without giving shape to some of my
endless, excess dream.
There is no loneliness until you know
yourself as separate even from yourself;
then it starts to burn. I wanted something,
   without knowing what it was, and so in a sheer
burst of vast and green imagining I
threw worlds out, millions of them, seeds;
I cast myself adrift. I wanted to feel something
spin, wanted an axis, wanted to plant
myself and find out how to grow. It was
quite a vision, blasting myself apart,
bursting into the cool nothing of what
would, any timeless second, be space, all
set in motion in the blink of my non-

existent eye. And naive, still. I look
back now and it all seems so obvious,
clear, but of course it always does, later, in
retrospect. Quite a gift, the glance behind,
allowing us regret. Once I saw myself there
was no other way. Such fierce and tender
regard demands action, expression;
I wanted to be seen, but needed form. So
out I spun, propelled by some internal
force, throwing off layer after
layer of what if, universe after universe
posed just so, wrapping me around
myself even as I came undone, great
shuddering waves, passion born then
when I uncoiled, opened myself
panoramic and plunged into my own
abyss. Oh, and what could I do then,
caught firm in my own embrace, twined
in my own limbs, my own bliss, held
fast and eternal. I knew I was beautiful.
There were stars.
I slept, dreamless and more, and watched
myself be born into multitudes. When I
woke, I woke to color bathed in color, all
tossed across my breath like jewels
scattered for birds. All potent and wild
it was, all amazing, all that I was,
and with what savage grace, how
much, each contradiction, each snowflake
its own temptation, its own endless dream.

Slightly edited for space, this poem is the work of my daughter Leigh Gray Kenyon, who was the baby born during my NDE.

CHAPTER 11.

IS THIS REAL

People are always asking, "Are NDEs real?" For several years my answer has been, they are absolutely real as *experiences.* Now I say the same about us: *We are real as experiences,* as bundles of experiences. Our visible aspect is real in the physical dimension but will ultimately disappear, like electrons, into the unseen, spiritual dimension, where, it seems, we are more like light. At least, that is my present interpretation of the science and numinous evidence.

REALITY AS NON-OBJECTIVE

I have come to accept that what the Yin/Yang circles told me in my NDE was true: We are not real in the way we originally mean it; we are, in that substantive sense, illusions. In quantum physics terms, as in spiritual reality, we have lost the solidity we believed our substance gave us; we are not real. Yet persistence has paid off, for beyond that literal fact is another truth which says that does not matter; we are always real.

Beyond that, I say that through those experiences we encounter both light and darkness, and that denial of the dark emotions is fear talking, and that refusal to face what we fear keeps us locked down. Especially if we say we are governed by love, we are called to give up our devotion to divisiveness, to do our inner work, to look our demons in the eye and walk through them to wholeness. That is integration. There is so much *beyond* the demons!

Yet we are still human, with our human blind spots. When I read a Buddhist claim of illusion, I still feel the old chill and depression descend, and the sureness of my new thinking wavers. For this reason, I pick my battles. Although I worked with a Buddhist teacher for a decade and learned Vipassana meditation, I continue to live my now-nontraditional religious life in a Christian setting which accepts a range of beliefs (even so, not always a comfortable fit; I am working on being less resentful of the discomfort). Those readers who are familiar with Theosophy will recognize a tilt in my comments toward the universality of the Wisdom tradition and its basis in nature, though I am not a theosophist.

As for reality, it pleases me beyond measure to include here as one of my expert witnesses RuPaul, the drag queen legend. Behind all the dramatic flair is a thoughtful, brilliant man. He is another representative of what I see in the new cosmology, an unexpected voice as authority. Here, in an excerpt from an interview with *Vanity Fair* magazine, he talks about drag's being all an illusion, building another reality:

> A superficial aspect of drag is mainstream. Like, the 'Ooh, girl' or 'Hey girlfriend!' or 'Yaaas.' That's mainstream culture," he says. "But true drag really will never be mainstream. Because true drag has to do with seeing that this world is an illusion, and that everything that you say you are and everything it says that you are on your driver's license, it's all an illusion. Most people will never in their lives understand what that is. Because they don't have the operating system to understand that duality.
>
>   I love that scene in The Matrix where you see the countless [rows] of people living their lives in a pod, but they're dreaming about this other world. That is such a powerful picture. I think most people have the ability to understand that but dare not go

there. Because then they would be forced to deconstruct their whole belief system and build another one. Building a new belief system and then maintaining it is a tall order. A lot of times it means you have to leave your family and friends behind. Because they're not gonna get it.[i]

Sometimes, to find (the truth of) yourself, you have to leave loved ones behind. If that isn't an echo of Jesus! And if that isn't straight talk for near-death experiencers wrestling with their post-NDE lives, I don't know what would qualify. The experience shows us marvels and/or terrors; but then we may be required to tear apart the dailiness we have lived in to restructure a new one. Sometimes what looks like illusion holds greater reality. Leave illusion to construct a new, illusive reality. If we do not do that, we will have no peace.

METAPHOR

The Abrahamic tradition has always taken the physical world seriously. Israel built its religious life around the relationship with God and land and a book; Christianity built its religious life around a relationship with God, an incarnation, and a bigger version of the book. Tangibility has always counted, and relationship has always been central! So it is no small thing, to say we no longer have the luxury of tangibility. I know, it seems impossible; but this is our new reality: that in the world of the twenty-first century we are not real in the ways we used to think we were. The Bible chimes in, from James 4:14: "We are but a vapor." The secret, it seems, is to discover more about where we are, and to learn new ways of thinking. We need to learn to live by metaphor.

> John Ryan Haule warns that we "become literalists when we think that our own journey has provided us with a 'gospel' to preach to others. Our journeyer discovers only its own 'good news.' … To take the imagery of the journeyer's *sight* in a literal manner is to get stuck in another broom-closet. The imagery

of each sojourn constitutes an opportunity which is also a dangerous challenge. The opportunity lies in escaping from the broom closet.[ii]

The new reality is that we are larger than what we have thought was our life, and far larger than its ego; everything really is all one. [iii]

All of us are exploring and changes come swiftly. No one has hard and fast answers about how the world and the universe work. We need to learn how to read life with the flexibility of metaphors, as models, as representations. The key is *how we think about things, about the world and meaning.* It is possible to avoid rigidity by assuming an underlying "it is as if…"

IDENTITY AND BERNARDO KASTRUP

When I first discovered Bernardo Kastrup, he was doing his PhD in computer engineering, writing so clearly and with such vision about artificial intelligence, I was immediately a fan. Next it was quantum physics, even better. And here he is in a *Scientific American* article from 2018 that I hope you will track down and read. All quotes here are from the paper, and its URL is in References.

The article is titled, "Could Multiple Personality Disorder Explain Life, the Universe, and Everything?"[iv] It's quite short and beautifully (clearly) written. The three co-authors are Kastrup; Edward Kelly of UVA, who was lead author of *Irreducible Mind;* and Adam Crabtree, a Toronto clinician known for his work with dissociation.

"Multiple personality disorder" is the older term for what is now officially called "dissociative identity disorder." This is the condition in which a single individual appears to have more than one personality or, in other words, seems to be several distinct people. It has been commonly disbelieved, or at least considered likely fraudulent. Within the past five years, convincing clinical evidence has emerged, showing that the phenomenon is, in fact, both real and fascinating.

In DID, the person is not imagining what it would feel like to be someone else; that is what actors do. In DID, the person *is* the alters. As the authors here state, in extreme forms of dissociation, "the psyche gives rise to multiple, operationally separate centers of consciousness, each with its own private inner life." It is this creation of "multiple personalities," or "alters," which neuroimaging techniques have demonstrated to be real, with "an identifiable neural activity fingerprint."

Almost twenty years ago, I experienced being close to someone with MPD, and I attest that the literature is true. The person was a bright young woman recently diagnosed with schizophrenia. In fact, she (her primary personality) was one of seven or eight 'alters' linked with her body, only one of whom was clinically schizophrenic; the others showed no mental illness. In ordinary life, this meant that sometimes she was schizophrenic and sometimes not. Truly *not*; pretense was not an issue.

The alters, who were not all the same gender, ranged in age from maybe four years old to teens to adult; they had distinct personalities, voices, metabolisms, different favorite foods; a couple had allergies, though not to the same irritants. Sometimes one alter would have a cold or flu, while the others did not. They were, as this article's authors state, "operationally separate centers of consciousness."

What I found most compelling was the occasional emergence of a 'director' identity (not the primary personality) who knew at all times what was going on both biologically and psychologically with the entire group and particularly with the schizophrenic alter. Uncannily sensitive to body chemistry and brain function, this director alter would emerge to give clinically sophisticated information, alerting the therapist and others to a medication imbalance or impending psychotic episode. What does this suggest about the depths of our own psyches, our own potential for what seems an impossible level of awareness? The *Scientific American* authors say::

> Although we may be at a loss to know precisely how this creative process occurs (because it unfolds almost totally beyond the reach of self-reflective introspection) the clinical evidence nevertheless forces us to acknowledge something is happening that has important implications for our views about what is and is not possible in nature.

"What is and is not possible in nature"! By now you will recognize why this article has laid such a claim to my attention. That is the question I have been asking all along: What is possible?

Kastrup had recently published a paper advancing the view that "dissociation can offer a solution to a critical problem in our current understanding of the nature of reality." (This is why I so enjoy reading these people, that they are unafraid to ask impossibly hard questions. Don't mess around, just go for the nature of reality! Sure, why not!)

His project was to tackle what is known as *the hard problem of consciousness*. The hard problem is that according to the basic tenets of physicalism, reality consists of "physical stuff outside and independent of mind," and mind should therefore be explainable by brain processes; but "arrangements of physical stuff" cannot explain our *subjective experience of qualities*. How does physical stuff turn into *feeling*?

To get around this problem, the authors say, some philosophers propose that all physical entities in nature, even subatomic particles, possess some very simple form of consciousness. Under this view, called "constitutive panpsychism," experience is built into all matter, "not just when it arranges itself in the form of brains," and our own consciousness is then (allegedly) a composite of "the subjective inner lives of the countless physical particles that make up our nervous system."

Now, however, there is another problem: "There is arguably no coherent, non-magical way in which lower-level subjective points of view—such as those of subatomic particles or neurons in the brain, if they have these points of view—could combine to form higher-level subjective points of view, such as yours and ours."

The obvious solution to this physically insoluble problem is to assume that consciousness, unlike matter, is all of one piece, extending to the entire fabric of spacetime rather than being limited to the bounds of subatomic particles. The view is called by modern philosophy "cosmopsychism," of which the authors say,

> "Our preferred formulation of it boils down to what has classically been called 'idealism'—that there is only one, universal, consciousness. The physical universe *as a whole* is the extrinsic appearance of universal inner life, just as a living brain and body are the extrinsic appearance of a person's inner life. You may have spotted the next problem: If all of the universe is open consciousness itself, we ought to be able to read each other's minds and know what is happening everywhere all the time. There needs to be an explanation, at least a basis for one, to account for the fact that people have *private, separate* fields of experience. There must be an explanation of how a single, universal consciousness results in "multiple, private but concurrently conscious centers of cognition, each with a distinct personality and sense of identity.
>
> The authors take a deep breath:"And here is where dissociation comes in. We know empirically from DID that consciousness can give rise to many operationally distinct centers of concurrent experience, each with its own personality and sense of identity. Therefore, if something analogous to DID happens at a universal level, the one universal consciousness could, as a result, give rise to many alters with private inner lives

like yours and ours. As such, we may all be alters—dissociated personalities—of universal consciousness.

Moreover, as we've seen earlier, there is something dissociative processes look like in the brain of a patient with DID. So, if some form of universal-level DID happens, the alters of universal consciousness must also have an extrinsic appearance. We posit that this appearance is life itself: metabolizing organisms are simply what universal-level dissociative processes look like.

In the *Journal of Consciousness Studies,* which precipitated the *Scientific American* article, Kastrup had concluded:

> It can be summarized as follows: there is only cosmic consciousness. We, as well as all other living organisms, are but dissociated alters of cosmic consciousness, surrounded by its thoughts. The inanimate world we see around us is the extrinsic appearance of these thoughts. The living organisms we share the world with are the extrinsic appearances of other dissociated alters.[v]

Put it all together. Keep going. "The world cannot be as I have constructed it; it is unimaginably different."

---

[i] Lawson, Richard. "RuPaul, the Philosopher Queen," *Vanity Fair,* November 2019, https://www.vanityfair.com/hollywood/2019/11/ rupaul-cover-story.
[ii] Haule, 142.
[iii] https://www.newscientist.com/article/mg24332410-300-is-reality-real-how-evolution-blinds-us-to-the-truth-about- the-world/#ixzz6CHGljgBE
[iv] Kastrup, Bernardo, Kelly, Edward; and Crabtree, Adam. https://blogs.scientificamerican.com/observations/could-multiple-personality-disorder-explain-life-the-universe-and-everything.
[v] Kastrup, Bernardo. *Journal of Consciousness Studies,* Volume 25, Numbers 5-6, 2018, pp. 125-155(31).

CHAPTER 12.

THE VOID: EL COLLIE AND STRANGE PLACES WITHOUT ANSWERS

The only person I have ever encountered with an experience much like mine died in 2002, thirteen years before I ever heard of her. The greatest difference between my NDE and the 'twin-like' experience of the woman named El Collie is our level of spiritual sophistication: I was like a preschooler with a box of crayons, while El was all vivid acrylics, her entire adult life a spiritual chaos overtaken by massive episodes of Kundalini involvement.

In 2000 she heard me interviewed on The Learning Channel, and wrote about the similarities of our experiences but did not contact me. She had spent years in the grip of Kundalini energies and had done prodigious research; she had a newsletter and blog, and a following, and knew infinitely more than I about how to read the experience and how to talk about it. Yet for all our differences, it was clear we shared a togetherness beyond logic or skill assessments.

Fortunately, much of her writing has been saved. In a blog post El quoted from my Learning Channel interview, referring to my "awakening." (Well, there was a first! Nobody'd ever said *that* before!) Here I was, speaking from the year 2000:

"There is a gift in these experiences. Now, it's not a gift we want to get, but if we're stubborn and hang in there, we work through a lot of issues. We come to discover our religious faith in incredibly deep ways that we couldn't if we just dazzled around on the happy level. So what I'm trying to do is go beyond the idea that pain equals bad equals punishment equals hell equals eternity equals despair. Because the alternative to despair I think is joy, which is different than happiness. But the paradoxical nature of this is that in order to get to real joy, we have to be able to accept suffering as part of us. And I know that sounds bizarre. But I didn't make up the rules. . . and it just seems to work that way."

For several years El had been working on a book which was never finished; the link to her saved work is in References.[i] The entire work reads like a master class in life with Kundalini. Here are some quotes from the section in which she muses about what was shared in our experience.

> Common ego-suspension experiences ... are very different in their psychological impact than the stark confrontation with the illusory nature of existence which Nancy Bush and I encountered. The positive experiences have a melting-quality whereby ego-boundaries are blurred and we feel ourselves to be One with life. By contrast, being divested of all previous notions of self is a great shock to the psyche. At this deepest level, not only one's sense of individuality but one's total sense of reality implodes. One's entire perceptual orientation is turned upside down and inside out...
> Awakening to the 'eternally complete consciousness' isn't about being in the presence of the One or feeling union with God, both of which assume the existence of two entities, self and Divine. In this experience, one's personal identity is obliterated. ...The collapse of the phenomenal world ... is disemboweling to the psyche. ...The problem was that what remained was a single Consciousness which existed in absolute aloneness.
> There is no one standing apart from the One to bear it witness when awakening occurs. ... Becoming Self-Realized is the experience of knowing there never was and never will be anyone to become enlightened, and that nothing but Consciousness IT-

Self is eternally real. ... Throughout my life this knowledge has followed me as a reminder that nothing in this world is entirely as it seems, particularly not my own ego-self.

Grof distinguishes between experiencing deities and divine personages ... and experiencing the core God/Self — the I-AM of pure consciousness. Many of the people who have this core experience ... seem to be exhilarated by the absolute freedom of realizing that everything and everyone is an illusion. But some – like Nancy, me, and others I've met who are more love-and-relationship oriented – are devastated by the eternal aloneness of Self/God.

... For a very long time, although I continued to function normally on the surface, I was in a twilight world where nothing, including myself, seemed to have any substance. I pretended not to know what I knew, and I was ever in search of an illumined soul who might somehow help me bear the weight of my secret knowledge.

Trying to come to terms with my lasting sense that nothing was real, I went on a rampage of reading ... in hopes that I might come across some piece of wisdom that would rescue me from the immensity of what I knew. ...

Most of the authors of the spiritual texts who described the God/Self realization were exultant and bubbling with promises of eternal bliss. Almost nowhere was there acknowledgment of the devastating part of the experience. ... Yet the irony was clear: the only ones able to understand what was being warned against were those who were already too far into the journey to turn back.

I had a realization that joy and suffering aren't opposites, but balancing sides – eliminate either side and the cosmic seesaw is broken, the whole momentum is lost. ... Love is supreme; it is the sacred life-giver of the universe. I don't think it matters much what level of revelation/epiphany/ awakening we've had – If we don't get that love is the key, we're still blind. And if we get just that much, we're blessed.

A reader echoes the reaction of almost everyone when he exclaims, "I fail to see what is so lovely about the thought that everyone I love doesn't really exist and I'm doomed to eternal loneliness." That is, of course, the key issue in resolving this experience.

The size and scope of the dilemma rises behind another experiencer who writes, "The undeniable thing about the Void is that it just won't go away—Nothing, no amount of faith or reason or denial, can budge it. Once it has 'known you', you are its for good. Thus, reconciliation seems the only viable way to me."

And yet, El specifically acknowledges that some people "who have this core experience (which a friend of mine calls 'God-in-the- Void') seem to be exhilarated by the absolute freedom of realizing that everything and everyone is an illusion."

She did not find that; nor have I. No wonder the integration has taken so long.

When I first read her words about the being-with-Being, it gave me such a shock of recognition—*yes! of course!, oh, that's it!* That second, nothing-sounding part of my experience has, for me, such a sense of enormity about it; but that's when I go wordless and can't think. El, with her years of Kundalini experience, was infinitely more articulate than I in every aspect of this communication. Without her words, I would be nowhere.

Here is what El Collie and I mean about the Void, from a website she liked: 'Hermetic Philosophy and the Mystery of Being,' by Alice Ouzounian: [ii]

> What is Spiritual Emptiness, or the Void? To answer this, we must first find out if it is possible to discover its essence within our being. Since the essence of Being is complete peace and stillness without reflection or any kind of manifestation or projection, could the essence of Being in of itself be called *Emptiness*? We must turn to the language of symbols to convey what it means, since Emptiness is a level of mind and consciousness, a level of realization that belongs to all traditions.

For example, in the *Vajra*, a ritualistic instrument used in Tantric Buddhism, Emptiness is represented by the central dot and symbolizes a focal point, the seed of the spirit in which everything is in a potential but static state. It also characterizes the *central axis* and heart of the universe. *Vajra* literally means the diamond scepter, or the thunderbolt.

Emptiness is also linked to the creative Void, meaning that it is a state of complete receptivity and perfect enlightenment. It represents the disappearance of, or more precisely the merging of, the ego with its own essence, which in Western tradition is called the Universal Soul or the Unknown God, and in Buddhism is called the Clear Light.

This merging of ego with its source could also be explained as the experience of infinite space, a frame of mind that confers us with the realization of the interdependence of all phenomena in creation, an experience that bestows an incredible degree of being true to ourselves, in that when our ego merges with its source, truth becomes impersonal,

However, as the letters of the alphabet are "tools" to build and interpret the written word, realities are the tools that our psyche uses to interpret and understand the world around it. Our psyche translates and perceives realities as *symbols*, but since this process of reading and understanding realities takes place in a world of duality, then realities belong to the world of the ego, which interprets and understands them. Hence, realities have nothing to do with the level of consciousness called Emptiness or the Void, since in the Void, there are no manifestations.

WHERE IT ALL COMES DOWN

The world of the circles

The first section of my experience, with the circles, has received the greater amount of attention, which makes sense because it is the part containing *things*. That's what we know to relate to (or against) in the world of Ego, as Ouzanian says. Their message was shocking and painful. At the time, it filled me less with dread than with grief, though after waking those emotions reversed.

It was the old religious tenets, the judgment, punishment, hell—even though they were not part of my own upbringing—which kept me locked in at first, whether I believed them or not. They were like a cage built of the only meanings I knew of. Back in 1962, as soon as I rejected the notion of predestination, the Stage 3 level of faith simply began to evaporate; I could no longer accept as given many of the arguments of Christianity's traditional theology. After the ferment of the first twenty years, and then in the spiritual encounters described in these pages, I realized that somehow, I had reached a different place.

As I marinated in other people's love-filled NDEs, my religious life shifted (and has kept doing so).

- Through pea-shoot awakenings and bent spoons and innumerable synchronicities.

- Through being unexpectedly present as observer at a startling physical healing with Joyce Hawkes and a guest healer, I have seen and felt under my hands the physical actuality of a healing. It is now an intellectual given.

- Through a healing circle experience at a conference in Kutztown, PA, I have *a physical memory of* major energy. I had been assigned to be one of the healers and had objected that as a novice, I had no idea how to do that. I was panic-stricken at being found out as a fraud; but the leaders assured me all selections were made through prayer; so there I was, standing behind one of the healing chairs, ready to imitate whatever could be seen of what the actual healers did. I've never prayed harder than in that circle! At the end of the public ceremony, the leader who was my mentor came to my chair…and sat down! There was nothing to do but submit to what I was sure would be my humiliation, as she would surely know me for a pretender; yet I went through the motions, and afterwards she came around the chair and put her arms around me. It was as if Old Faithful had just erupted beneath our feet, carrying us with it—a gigantic upwards whoosh of light and force. There is no way I can deny or forget the truth and power of that energy, though I do not generate anything like it myself. (Less inspiringly, on the day of a maximum-stress IANDS Board meeting, with its future at stake, the main gasket on two unrelated cars blew as I was driving them.)

- Mysteries have been building up, all the things I was learning which were true, and it hasn't seemed to matter that I didn't understand them as long as I could accept them as real. (You see a growing point here?)

About the compassion...

There is a boundless literature about how coming close to death, or enduring great pain, or having some kinds of altered state experience increases a person's compassion. This doesn't necessarily mean more *emotionality*, more *sentiment*, or sending Hallmark cards. It is more about recognizing a shared humanness. After the experience, people have a deeper sense of awareness of the reality of others, of knowing their fragility and wanting to serve their being.

About the joy...

The joy starts with a repeat quote from the previous chapter: If there is an eleventh Commandment, it must be, *"Thou shalt do thy inner work."* The agony of the inner work comes first. Astrologically, I am a strong Aries and have had to battle through shadow work and accept submission to Spirit. For extended periods, that struggle has been agonizing. Yet that acceptance is key; we have to acknowledge suffering as well as pleasure. Out of the inner work, and the compassion, and the giving-over of self, I have discovered there is a very real joy, deeper than ordinary happiness. It is, I think, related to Being itself—like a flash of spectacular lining on a beautifully made garment; only a flicker, but it makes all the difference. The joy is not showy display but like a quality of light, subtly under everything, making all the difference.

About El's claim of exhilaration:

A reader of my blog wrote, "El said something that I think must have been a gross exaggeration—'Many of the people who have this core experience… seem to be exhilarated by the absolute freedom of realizing that everything and everyone is an illusion.' I find this unbelievable—of the relative few that I've read about or known who have had this "experience,' not one of them found it to be a positive experience."

No one I know of has claimed to find perceived illusoriness a freeing experience; however, this is a Western response. I have not posed the question to Buddhists. I am quite certain that is a very different approach.

THE YIN-YANG.

The same reader who asked about illusoriness also said, "I've never heard you expound on the obviously profound significance of this eastern symbol within your NDE. As impish as they were, they represented life: relativity, duality. And being dualistic, they were Janus-faced, betraying life's most profound paradox: Life as both savior and betrayer."[iii]

Life as both savior and betrayer. Ah.

Caught. I am indebted to readers in so many ways! Steve's comments and his brilliant similes drive straight to the heart of what makes NDEs important, far beyond their superficial "golly-gosh" phenomenology. Beneath their eye-widening images lie treasures, if only we can get to them. Sometimes we are reluctant to go looking.

In the most common usage, as I have seen it, the Yin-Yang is interpreted as a symbol of opposites in conflict. In the more useful and far more nuanced definition as described by Wikipedia, the symbol describes "how apparently opposite or contrary forces are actually complementary, interconnected, and interdependent in the natural world, and how they give rise to each other as they interrelate to one another. ...[They represent] complementary (rather than opposing) forces that interact to form a dynamic system in which the whole is greater than the assembled parts." That makes far better sense than simple opposition!

Scottish homeopathic physician Bob Leckridge expressed in a long-vanished blog post a common thought when he said, "I love what it represents, that flowing balance of darkness and light, the harmony of the male and female energies, and the subtle hint that each opposite contains the other." Nowadays, everyone pretty much smiles and agrees. Had my experience occurred a decade later, my response would no doubt have been different, but these and Steve's thoughtful interpretations were not what registered with me.

Although it is difficult to remember such an actuality, the Yin-Yang symbol was not instantly recognized throughout the United States in the early 1960s. From within mainstream Protestant culture of the time, I saw the images in my experience merely as unusual 'circles.' There was no familiarity with the symbol, which was not a part of my world, my upbringing, my understandings, and could bring nothing to my perceptions. As a college graduate, my optical nerves had no doubt *seen* the symbol, but without registering meaning; as a result, for many years its only resonance for me was not 'flowing balance' but the dread I associated with a message I took to mean universal annihilation. Otherwise, they were simply a blank.

The Yin-Yang as symbol remained submerged beneath the shock of the Void as experiential residue; it took a long time and a determined effort to be able to distinguish between the symbol and the experience. Ironically, the symbol was until recently the core of the IANDS logo.

One has to marvel at the elegance of its design, both visually and epistemologically. Beyond that, though, I now see the symbol rather as one might an old lover/rapist with whom one has a child, sharing simultaneously the memory of a painful encounter and the implacability of being locked into a continuing acquaintanceship, conducted with unavoidable caution, yet aware that the reality created new life and growth—itself a miracle.

Life as both savior and betrayer. Ah. A reality. I can see that now.

THE WORLD OF THE VOID

The second section of my NDE, without the circles, leaves me with a sensation that I am missing things—but of course, because *there are no things*—but also that meanings are eluding me. Then it occurs to me that this is what the ineffable *feels like*, a kind of palpable frustration full of what might be shadows if only I could see them.

It was El Collie's words which showed me what I had been trying to see, her understanding which gave me the beginning of a comprehension. She was able to articulate what I could not even begin to shape in my mind.

El's writings sent me to the site of metaphysician Alice Ouzounian and her center for hermetic spirituality on Cyprus. She says of the Void and emptiness:

"Has reality a connection with Emptiness? Reality changes all the time. It moves and is transformed incessantly by the interpretation and experiences of our individual ego. Reality is part of life as it brings the necessary changes and meanings that we all need to express ourselves. However, with the incessant changes that we all need, realities seem to change with the transformation of consciousness. It can never be fixed, never be constant, but remains limited in time and space. Reality is as fluid as water, as ephemeral as air, and therefore, in itself, can never be rooted, anchored or possessed, since it comprises nothing other than changing appearances. So clearly, since reality is a projection of our own understanding of something, Emptiness is devoid of reality.

"However, as the letters of the alphabet are "tools" to build and interpret the written word, realities are the tools that our psyche uses to interpret and understand the world around it. Our psyche translates and perceives realities as *symbols*, but since this process of reading and understanding realities takes place in a world of duality, then realities belong to the world of the ego, which interprets and understands them. Hence, realities have nothing to do with the level of consciousness called Emptiness or the Void, since in the Void, there are no manifestations.

"Emptiness designates a state of mind, an inner level of consciousness based upon the renunciation of what one believes to be real, beyond all comprehension or lack of comprehension. Emptiness is, therefore, a higher level of the mind, an attunement with "nothing," i.e., with Pure Being that has no reflection at all.

"…Returning to the question of what we mean by Emptiness, it is difficult to explain because it represents a purified level of consciousness in which objects and creatures are seen as transitory realities. Hence, we could say that Emptiness in itself is the source of Being and therefore *formless*. However, since it is the source of Being, it nevertheless nurtures all realities. Symbolically, we could say that Emptiness is the dark womb of creation."[iv]

Australian near-death experiencer Alex Paterson has described having several encounters with the Void as a youngster of seven or eight:

"The experience was terrifying to me as a child because the VOID had no actual form and as such was empty, yet paradoxically was so big as to be seemingly infinite (even though I couldn't conceptualise infinite at the time), hence my childhood perception that it was HUGE. But perhaps more disturbing to me as a child was that IT could not be defined in any terms of this reality in that it had no tangible presence even though I was intimately aware of its presence (a paradox). Thus, there was no sound, no sight, no taste, no smell, no feel and no emotion about it, and yet somehow I was fully aware of its presence. Because the VOID seemed to be everywhere yet paradoxically nowhere, there was nowhere for me as a child to hide from it, which was why I was so terrified of it. I might add IT never did anything to harm me; it appeared to be devoid of emotion and just seemed to be dispassionately observing me. (The quintessential dispassionate observer so to speak). …Having experienced the VOID, I am now able to grasp the scientific concept of the notion postulated by quantum physicist, Dr David Bohm, of an "implicate enfolded order which exists in an un-manifested state and which is the foundation upon which all manifest reality rests." [v]

Stanislav Grof's holotropic breathing therapy sometimes taps experiences which connect to the Void:

"When we encounter the Void, we feel that it is primordial emptiness of cosmic proportions and relevance. We become pure consciousness aware of this absolute nothingness; however at the same time we have a paradoxical sense of its essential fullness. ... While it does not contain anything in a concrete manifest form, it seems to comprise all of existence in a potential form. In this paradoxical way we can transcend the usual dichotomy between emptiness and form, or existence and non-existence. However, the possibility of such a resolution cannot be adequately conveyed in words; it has to be experienced to be understood."[vi]

El Collie's words were for me a breakthrough, as were Ouzounian's, because I have never been able to think about the Void consciously. Still unable. When I began seriously trying, I would come to my desk, full of focus and intention, put my hands on the keyboard, and find myself waking up fifteen or twenty minutes later with not a word typed, no memory of the gap, and nothing to say; I assumed this was at least in part some sort of post-trauma aphasic response, as if my head were muffled or I was being firmly but nicely silenced. Now I am not so sure.

British mystic Paul Brunton, whose Notebooks contains a lengthy section on the Void, observed that just at its edge, "Not only does the mind become utterly blank and lose all its thoughts, but it loses at last the oldest, the most familiar, and the strongest thought of all–the idea of the personal ego."[vii]

It all simply *goes*. I am now thinking that perhaps my blank-out reaction is simply being picked up and thrust for a short while back into the Void. There is no thought *about* the phenomenon; there is only the Void itself. It has no emotional content, no intellectual burden; it is simply *no.thing*. As with so very many topics, I do not know; it is a wonder if perhaps…

What We Don't Know Hurts Everyone: Misinformation

Whenever possible, in everything I write I include some statement that there is no research evidence that "good people get good NDEs and bad people get bad ones." My insistence on claiming lack of evidence has not endeared me to virtually everyone else who writes about distressing experiences. They seem to find my statement an opposition to the common wisdom idea expressed by thinkers such as Deepak Chopra:

> "Our thoughts being energy, it only makes sense that our repeated images, affirmations, visualizations, deeply held beliefs, fears and desires, vibrating within the larger web of reality, would have an effect upon that reality. In fact, when you stop and really think about it, since we are all connected, how could it be otherwise?"[viii]

Of course it makes sense! But not total sense. And making sense is not the same as having statistical evidence to back up the anecdotes, which is what we lack.

There is *abundant* anecdotal evidence in the accounts that people who report distressing NDEs (or their interviewers) believe their experience resulted from pre-existing meanness or a similar crippling pattern. However, to my knowledge, no study has ever tested those claims to see if what the individuals believe is actually true. Are they reporting accurately that they really were as hurtful, as antisocial, as out-of-the-norm dysfunctional as they say? Or could it be they believe they must have been really bad because their frightening NDE wouldn't have happened otherwise? And where their behaviors were *not* out of the norm, perhaps we can begin to kick the guilt-producing explanations to the curb?

It is well documented that people often evaluate their own weaknesses or misbehaviors more harshly than observers do, so their subjective views may be disputable. On a normative scale, where would they stand? And, of course, there is always the issue that no one has asked similar characterological questions about people with pleasant NDEs, which says a good deal about research attitudes.

We come from a reward/punishment culture which has only anecdotal data from people with reward/punishment upbringing. Of course they will support claims that previous (negative, sinful, unhealthy) attitudes determined having a distressing experience. We are still lacking *statistical data* to say whether that belief is supportable. Testing would tell whether the negative attitudes were the person's norm or merely a passing mood. It would tell whether the level of negativity was within normal range or at an extreme. These statistical details matter when talking to people who are trying to make sense of the whole issue!

Why is it important? Because when we do not take care to examine the details of what we believe, we lead ourselves (and others) astray. We have no basis for knowing what is truly the energizing force behind these experiences. Maybe it is the 'negative' emotions; but it could just as well be something more complex. Sloppy thinking is more than incidental; it is a flaw in the very foundation. This is why people like Kastrup and Kelly and Greyson are so vital: because they are always paying attention.

---

[i] El Collie. http://www.kundaliniawakeningsystems1.com/downloads/branded-by-the-spirit_by-el-collie.pdf.
[ii] Ouzanian, Alice. http://www.plotinus.com/spiritual emptiness.htm,
[iii] Steven Weber, private message.
[iv] Ouzounian, Void and Emptiness
[v] Paterson, Alex. http://www.vision.net.au/~apaterson/spiritual/ void.htm.
[vi] Grof, *Cosmic Game* in Paterson.

[vii] Brunton, Paul. *Notebooks*. https://paulbrunton.org/notebooks/23/8.
[viii] Chopra, *op cit.*

CHAPTER 13.

THE MAP IS NOT THE TERRITORY

The universe is our elephant. We try to describe that part of it we grasp. Although it can neither be taken in by a human mind nor described completely from any one perspective, that has not kept us from trying. From before the Bronze Age until today, no matter our place in geography or our status in culture—we have all been looking out on the same universe, looking to the limit of our senses, and interpreting our descriptions. From these models have developed the many human theories, doctrines, beliefs to create the systems we live by: all the forms of religion and types of spiritual discipline, the varieties of philosophy, the views of science and history, each believing itself to have some truth about the elephant of the universe.

There are those who claim this view of mine to indicate the end of faith, as it advances no single Truth. I would say my view is merely one exploration from the perspective of a single voyager, and it carries its own sustaining faith. For me, this understanding provides renewed faith, on a bedrock of unshakeable trust.

What struck me some years back was the stunning realization that, as different as people's belief models are, many display a remarkable underlying consistency in what they consider revealed truth. Then I discovered that other writers, chiefly Aldous Huxley, had arrived at this conclusion before me, and had given it the title Perennial Philosophy. Considering a group of these models together, one can trace their common characteristics. Each set of descriptions is a map of the great cosmic territory.

One model is made up of themes shared by sacred writings. These traditions acknowledge a "something other" in creation, a sensed holy presence, a powerful creating and shaping force; in English, it is often called "God." These scriptures of religious tradition include abundant mention of radiant light. They speak of the importance of caring, of loving one's neighbor, even one's enemy. And whether in stories of Eden or of lost Ancient Ones, they tell of an unimaginably distant time of harmony and wholeness.

A second model is present in the writings of the world's great mystics, those persons past and present who live with a radical sense of inbreaking spiritual experience. Their lives have been captivated by a sacred Presence, by an unseen Beloved. Their experience is described in terms of light and darkness. They are overwhelmed by the centrality of agape, of unconditional love. And they attempt to describe their flashing moments of union with the sacred Presence as a seamless whole.

The third model comes from the accounts of near-death experiences across the centuries. People who have had a near-death experience talk about light¾its positive presence or absolute absence. They tell of presences, sometimes of a sacred Presence they may call it by a name familiar to them from their religious tradition or refer to as a Being of Light. Their lives afterward are driven by the conviction that the most important thing is love, that everything is connected, that there is a oneness to everything in the universe.

From a fourth and quite different realm come the models of quantum physicists, their science described not so much in words as in mathematics. They have demonstrated that the most basic bit of existence is the photon, the smallest particle of light. Increasingly they say there seems to be, somehow, a shaping intelligence behind (or within) the workings of everything that is. They speak of their search for a unified field theory, following hints of a theoretical commonality linking all things. And particle physicists have documented the unarguable truth that at the sub-atomic level, there can be no objective observer and observed, for in a mysterious dance of oneness everything at the quantum level interconnects and interacts.

It would not be true to say that because these models share some commonalities, they are all saying the same things. To claim so would be no less naive and uninformed than to insist that all religions are alike simply because they all deal with the spiritual. Physics is most emphatically not like religion, nor NDEs like disciplined mystical tradition. They are not all precisely the same ... but *because they are all describing the same observed universe, they point in a common direction.* Taken singly, the models may have wildly differing interpretations; seen together, they describe a pattern in the universe: A mysterious and powerful motive force, perhaps a shaping consciousness. Light. Unity. Interconnection. Relationship.

THE NATURE OF THINGS

Many years ago, the same sister with whom I had shared the linoleum rug was spending an extended time with my marriage family. We were in the kitchen one afternoon—Babs and I, her five children and my three, aged from two to perhaps eleven. The kids crowded in around the big round oak table, playing some thunderous game that involved smacking the surface,. After a time, one of the littlest ones asked,

"What makes the noise?"

This was a group never at a loss for theorizing, and they came to sudden attention. My nephew, Paul, who was nine, could always be depended on.

"Noise," he explained solemnly, "is the sound of molecules screaming. They're very small, and when something hits them, they explode. So they scream."

Everyone stared intently at the heavy table. Then:

"This table is really old. How come," asked a six-year-old, "if the molecules have been exploding… how come the table is still here?"

Babs and I did not dare look at each other but waited for Paul's response. It was bound to be good.

"Well," he said, "the table is still here because . . . because molecules are always pregnant, and when they explode the babies escape, so there are always new molecules."

My eleven-year-old was exasperated. "That is so stupid! How can molecules be pregnant?"

Paul looked at her with cool assurance.

"Because," he said, "it is *in their nature* to be so."

The Hebrew word for 'angel' is *messenger*. What I believe about near-death experiences is that, like Paul's molecules, it is in their nature to be a kind of angel, bringing messages similar to those that founded our religions. They are also much like the findings of contemporary physics and the testimony of humanity's mystics, clues to natural truth. I believe these altered states of consciousness are fleeting excursions beyond all maps, into the Territory itself¾momentary trips inside the skin of the elephant, experiential blips into the nature of the universe. As Rudolph Otto knew, sometimes the blips are terrifying to human minds. That, too, is part of the universe.

Of course the experience is ineffable, because All-of-Creation cannot be encompassed by the mind of a single human, nor can it be squeezed into syntax and grammar and vocabulary. The experiences announce that it is in the nature of the universe to be a dance of consciousness, of radiant light and horrifying darkness, of all things in relationship, and of a mysterious and ultimate unity. And because we are children of that universe, it is in *our* nature to be so as well.

I have come to see that my NDE is the farthest thing from negative. It is intricate, subtle, cross-cultural, interdisciplinary. Along with taking me through a dark wilderness, it has led me to confidence in the affirming beautiful NDEs, which helped me lead my mother to her peaceful death. I say with Mishka Jambor that even painful NDEs need not be negative; and with Miriam Greenspan, that 'personal odysseys through the dark emotions [can be seen] as a path to sacred power.' I say with El Collie, "This was the most harrowing, soul-shattering, and simultaneously the most illuminating and transcendent experience of my life."

I believe that NDE accounts, like our sciences and our theologies, offer glimpses of the nature of the universe, of the nature of God, the Sacred, the Source—choose your terminology–and that, radiant or terrifying, they illuminate our understanding of where we are and what is required if we are to live well in this place:

> *The Sacred is in your midst. Pay attention. Love what is holy. Care for each other and for the world. Be just and merciful.*

FINALE

The last words here are an anecdote and another poem. When I was still living in New York City, back in the days of the anthropological library, my dream was to become a Broadway lyricist. Instead, I married and moved out of the City, and life went on. Years and years later, there came a phone call from the music committee of that big and wealthy church I had joined. The church was well known for its spectacular Boar's Head festival at Christmas, and they were thinking of balancing that with something around Easter. They didn't know quite why they were asking me, but they knew I did some writing, and would I be interested in working with a composer to come up with a few songs?

And an invisible window opens, and an invisible Spirit reaches in...

The composer, Toby Hall, was an insanely talented pianist, newly arrived from New York. He had been a rehearsal pianist at the Met, had worked with Sondheim, and had moved to Hartford to be near his sister. He was slowly dying of Hep C and AIDS.

We were the unlikeliest partnership. He was 41 and I was almost 60. He was Manhattan through and through, and, as my husband used to say, was "gay as a tree-full of larks," which meant a particular type of paradoxical antipathy for women. He had a wicked sense of humor; the acid of his tongue could leave scars. He had grown up with family dysfunction and had huge Mother Issues. I, only a few years younger than his mother, had long since lost my Manhattan edge, and had spent many years since my twenties struggling, as a woman, to maintain any sense of identity in circles of gay men, my husband among them; I also had Issues as big as a barn. But I wanted to write lyrics and he wanted to compose, and we were both particular about quality.

The partnership was better than a success; someone once laughed that we were, in a way-offbeat way, like Twin Flames. (Here is the heart of the suffering issue: to look the deepest and most painful issue in the eye and, rather than running away, to walk straight into it and learn to come out the other side. Through the work, we walked each other through our demons.) The music committee got more than "a few songs": it got anthems and a song cycle and a glorious little concert piece about the aurora borealis. They also got *The Eighth Day*, a cantata based on the ministry of Jesus, with Jesus written as a mime juggler and Satan as "the Razzle Dazzle Man," a Trickster magician.

What Toby and I got was each other, and resolution of some old Issues, and a musical triumph which gave us each a great sense of completion. We were able to write Toby's death and to express a shared view of theology not often said aloud in churches. Toby remained well enough to be at the piano for the first production. My mother was able to attend. For the rest of the story, see the closing note in Acknowledgments.

The Alleluia is the finale of the cantata *The Eighth Day*, music by Toby Hall, book and lyrics by Nancy Evans Bush, produced at Asylum Hill Congregational Church, Hartford, Connecticut in 1992 and 1994. (I wish you could hear the music!)

ALLELUIA

All of God's creation, born of incarnation,
all the dance of God.
Mystery of seeing: all creation being
light within the light of God.
God rearranging our life in its changing;
nothing is wasted or lost.
God in us rising, creating and prizing,
caring and sharing the cost.
God in the mirror—all coming clearer—
God is alive in our will;
Though body scatter, nothing of matter ever can be still.
Universe turning, light for holy darkness yearning
as the darkness yearns for light.

All coming round in the merciful wheeling
allness and wholeness, God's Life.
God in the dancing of atoms is chancing leaps we cannot see,
Particles whirling, galaxies curling within you and me.
Dying, living; rising, giving;
the world is the body and dancing of God,
and we dance in the darkness; we dance to the sun
and back to the darkness where dancing's begun.
    All in God's being we live beyond seeing,
and in our sleeping, still in God's keeping,
and dancing: Alleluia!
All of the world is the body of God:
energy dancing chaos to joy,
and it rises in wheeling and spins into form,
living and dying in light and in storm.
All of creation in bright transformation,
and in believing, moving past grieving,
and dancing: Alleluia!
We are the body and dancing of God;
and we dance in the darkness, we dance to the sun
and back to the darkness where dancing's begun—
all in God's being, we live beyond seeing,
and in our sleeping, still in God's keeping,
and dancing: Alleluia!
    All shall be well. And all shall be well.
All manner of things shall be well.
Alleluia! Alleluia!
Alleluia!

ACKNOWLEDGMENTS

After trying unsuccessfully to write this book for more than ten years, I was gifted by Covid-19 with a landscape of open space and time so vast that it has been possible to concentrate for weeks at a time without interruption. Once I was committed to the project and sitting at the computer for hours every day, information began showing up. In March, when this began, my NDE was not yet fully integrated; by September, it had found itself.

So, my first acknowledgment is to God—Universe, Source, Force, whatever word you use—my gratitude for a life that works this way and keeps surprising in astonishing ways.

The fact of the pandemic has imposed unusual conditions. Because this household is vulnerable because of age and health histories, we have been scrupulous about sheltering in place; there has been no in-person interviewing, no visits with experts. With libraries unavailable, my resource has consisted of whatever is on my bookshelves or available online.

These restraints have shaped the book in an unplanned but interesting way. The references are heavily Google-linked to authors who are not behind paywalls; in other words, they have been published not in the locked-away academic journals but in work-around, open access sources. The references are often blazingly contemporary (sometimes within days publication) and informal. There is a great deal of information here but not a lot of academy. I am grateful for the informality.

I am intensely grateful to the beta readers, David Maginley in Nova Scotia, Mark Anthony in Florida, and Steve Weber in California, who have given not only their differing perspectives but abundant good humor. Also in California, Susan Pomeroy, the Web Geographer, has designed the covers of all three of my books, maintaining both the *dancingpastthedark* website and my sanity for years; she has become a treasured friend. Lee Campbell in Seattle, is a helpful beta reader and proofreader. Enthusiastic thanks to chaplain Debi Wacker of Lynnhaven Congregational Church, United Church of Christ, in Virginia Beach, VA, for introducing me to *Integral Christianity,* without which I would have been hard pressed to know where this book was going. Special thanks go to my daughter Leigh Grey Kenyon, for her poem and a masterful job of editing. As always, my boundless gratitude to Nancy Poe Fleming for her thoughtful assessment of every draft and for all the sandwiches. Her support and smiling presence over the past 42 years have made this entire voyage possible.

Note: The end of the Eighth Day story.

Toby and I got two years of collaboration. In 1993, he and my beloved mother died in the same week, and by happenstance their funerals were scheduled an hour apart, four blocks apart, on the same afternoon. I would have to leave with the benediction of Mom's service and race to Toby's to deliver his eulogy. It was clearly an emotional impossibility, but there it was, and for those two "show must go on" people, there could be no excuses.

The day before the funerals, a cold November day, I was pacing around in my kitchen, railing at God and Universe about the unfairness of the situation. I was so wrenched by grief, surely this was too much to ask, an absurdity! The expectation was really pushing it, simply demanding too much! How could I even speak, much less deliver a eulogy? How unfair! I was storming, a tower of misery and hostility.

In the midst of my rant, there was a movement at the side window. I did a double-take; there was a bluebird! I had longed for bluebirds in that Connecticut yard for years but had never seen one, especially not in November. Toby had a little stained-glass bluebird over his desk. But there it was, a live one, and there very shortly was another one, a female. (Aw, c'mon! Really!) They flew from the window to the nearby lilac bush and back, and when at the window they would look inside. (I looked back at them. NO, it's too much to ask!) But they kept it up for more than an hour. (You've got to be kidding! You're just bluebirds!) Eventually, of course, I gave in and shouted, "Oh, ALL RIGHT!" and slammed my coffee cup on the counter. And the bluebirds looked at me and stayed, not always together but never far apart, all afternoon, until darkness fell. We never saw them again. And the next day I left my mother's crowded funeral service at the benediction and rushed to Toby's, where the soloist was poised to sing extra verses until I could arrive to deliver the eulogy. The show went on.

This is true synchrony, when, yes, it may of course have been pure coincidence that a pair of bluebirds would be migrating through that particular stretch of woods that particular raw November day, and would behave in that particular way. But given what I have discovered in "All This," it is equally possible and probably more likely that the bluebirds were not quite a coincidence. Whatever, or whoever, I am immeasurably grateful for their presence and for that possibility. Toby's stained-glass bluebird now hangs over my desk.

grief, surely this was too much to ask, an absurdity! The expectation was really pushing it, simply demanding too much! How could I even speak, much less deliver a eulogy? How unfair! I was storming, a tower of misery and hostility.

In the midst of my rant, there was a movement at the side window. I did a double-take; there was a bluebird! I had longed for bluebirds in that Connecticut yard for years but had never seen one, especially not in November. Toby had a little stained-glass bluebird over his desk. But there it was, a live one, and there very shortly was another one, a female. (Aw, c'mon! Really!) They flew from the window to the nearby lilac bush and back, and when at the window they would look inside. (I looked back at them. NO, it's too much to ask!) But they kept it up for more than an hour. (You've got to be kidding! You're just bluebirds!) Eventually, of course, I gave in and shouted, "Oh, ALL RIGHT!" and slammed my coffee cup on the counter. And the bluebirds looked at me and stayed, not always together but never far apart, all afternoon, until darkness fell. We never saw them again. And the next day I left my mother's crowded funeral service at the benediction and rushed to Toby's, where the soloist was poised to sing extra verses until I could arrive to deliver the eulogy. The show went on.

This is true synchrony, when, yes, it may of course have been pure coincidence that a pair of bluebirds would be migrating through that particular stretch of woods that particular raw November day, and would behave in that particular way. But given what I have discovered in "All This," it is equally possible and probably more likely that the bluebirds were not quite a coincidence. Whatever, or whoever, I am immeasurably grateful for their presence and for that possibility. Toby's stained-glass bluebird now hangs over my desk.

APPENDIX: RECOMMENDED READING

The dozen titles below point to significant books of my early journey. There were others as well, but these are all classics and still in print. They are of a vintage which does not assume the reader has a lot of background information. Highly recommended.

Fritjof Capra, *The Tao of Physics*
Edward Edinger, *The Creation of Consciousness*
Stanislav Grof, *Realms of the Human Unconscious: Observations from LSD Research*
Carl Jung, *Memories, Dreams, Reflections*
Lawrence LeShan, *The Medium, the Mystic, and the Physicist: Toward a General Theory of the Paranormal*
Caroline Myss, *Anatomy of the Spirit*
Robert E. Ornstein, *The Psychology of Consciousness*
Huston Smith, *Forgotten Truth: The Common Vision of the World's Religions*
Charles Tart, *States of Consciousness*
Tibetan Book of the Dead, The
Ken Wilber, *Up from Eden: A Transpersonal View of Human Evolution*
Gary Zukav, *The Dancing Wu Li Masters: An Overview of the New Physics*

BIBLIOGRAPHY

Brueggemann, Walter. *Message Of The Psalms.* Minneapolis, MN: Augsburg Fortress, 1985.

Bush, Nancy Evans. *Dancing Past the Dark: Distressing Near-Death Experiences.* Cleveland, TN: Parson's Porch Books, 2012.

___*The Buddha in Hell and Other Alarms: Perspectives on Distressing Near-Death Experiences.* Self-published, 2016.

Calone, David Stephen. *The Spiritual Imagination of The Beats,* Cambridge University Press, 2017.

Campbell, Joseph. *The Hero with a Thousand Faces.* Princeton, NJ: Princeton University Press, 1968.

Chodron, Pema. *When Things Fall Apart: Heart Advice for Difficult Times.* Boulder, CO: Shambhala. 2016.

Corbett, Lionel. *The Religious Function of the Psyche.* London: Routledge, 1996

Eliade, *Rites and Symbols of Initiation: The Mysteries of Birth and Rebirth* New York: Harper and Row, 1958.

Gebser, Jean. *The Ever-Present Origin.* Athens, Ohio: Ohio University Press, 1986.

Glucklich, Ariel. *Sacred Pain: Hurting the Body for the Sake of the Soul.*

Grof, Stanislav. *Realms of the Human Unconscious: Observations from LSD Research.* New York: Viking, London: Souvenir Press. 1975, 1993.

___ *The Ultimate Journey: Consciousness and the Mystery of Death.* Ben Lomond, CA: MAPS, 2006.

Haule, John Ryan. *Perils of the Soul: Ancient Wisdom and the New Age.* York Beach, Maine: Weiser.

Kellehear, Allan. "Census of Non-Western Near-Death Experiences to 2005: Observations and Critical Reflections," *Handbook of Near-Death Experiences.* Santa Barbara, CA: Praeger Publishers, 2009.

Lukeman, Alex. *What Your Dreams Can Teach You.* Audiobook.

___ *Nightmares: How to Make Sense of Your Darkest Dreams.* New York: M.Evans, 2000.

Maginley, David. *Beyond Surviving: Cancer and Your Spiritual Journey.* Halifax, NS, Canada: Tristan Press, 2016.

Peck, M. Scott. *The Different Drum.* New York: Touchstone, 1998.

Peterson, Jordan B. *Maps Of Meaning: The Architecture Of Belief,* Abingdon, UK: Taylor and Francis Books, 1999.

Peterson, Robert, Charles Tart. *Out-of-Body Experiences: How to Have Them and What to Expect.* Newburyport, MA: Hampton Roads Publishing, 2013.

Schäfer, Lothar, foreword by Deepak Chopra. *Infinite Potential: What Quantum Physics Reveals about How We Should Live.* New York: Deepak Chopra Books/Random House, 2013.

Sutherland, Cherie. *Transformed by the Light.* New York: Bantam, 1992.

Tarnas, Richard. *Cosmos and Psyche.* New York: Plume, 2006.

___ *The Passion of the Western Mind: Understanding the Ideas That Have Shaped Our World View.* New York: Ballantine Books, 1991.

Tart, Charles T. *States of Consciousness.* New York, E.P.Dutton, 1975.

Wilber, Ken. *Up from Eden: A Transpersonal View of Human Evolution.* Boulder, CO: Shambhala, 1983.

Zweig, Connie. *Meeting the Shadow: A Consciousness Reader.* Los Angeles, CA: Tarcher. 1991, *xiv.*

If you enjoyed reading *Reckoning*, won't you please leave a quick review at Amazon or your favorite review site? You could also tell a friend, even mention the book on your favorite social media. With these small publishing projects, your speaking up makes a big difference. Thank you!

Made in United States
North Haven, CT
03 April 2023